GENETIC ENTANGLEMENT OF THE PERSONALITY TRAITS OF SANGUINITY AND AGGRESSION IN CASE-CONTROL STUDIES

GENETIC ENTANGLEMENT OF THE PERSONALITY TRAITS OF SANGUINITY AND AGGRESSION IN CASE-CONTROL STUDIES

A.M. BENIS, Sc.D., M.D.

A.M. BENIS
New York

© 2018 A.M. Benis

1987654321

KDP Edition

ISBN 978-1-790-89054-5

Internet site: npatheory.com

All rights reserved under International and Pan-American Copyright Conventions. No part of this publication may be reproduced, stored in any information retrieval system, photocopied, recorded, or transmitted by any means whatsoever without the express written consent of the author.

We are standing at the threshold of an era in which the entire proud edifice of medicine — including psychiatry as a whole as well as psychoanalysis — will rest on genetics.

— Sandor Rado (1961)

ABSTRACT

Our objective was to examine how interacting genetic traits can affect the interpretation of case-control studies. The analysis is based on a model of personality reposing on three genetic traits: sanguinity (N), perfectionism (P) and aggression (A). The model is based on classical genetics, with N and A being high frequency recessive traits, and trait P being dominant. The genes for traits N and A are *entangled*, in the sense that if one of the traits is absent, then the other must be present. This entanglement of "complementary genes" causes infertility in certain combinations of NPA types. We examined hypothetical case-control studies with the use of Hardy-Weinberg methodology and gene frequencies to generate representative populations. Six different hypothetical case-control studies were replicated for seven different populations. The computed results of each case-control study were highly dependent on the genetic composition of the representative population. When the condition present in the "case group" was genetically related to either trait N or A, there was a reciprocal difference induced in the other trait due to their entanglement, leading to possible confusion in interpretation of the results.

CONTENTS

ABSTRACT   vi

PREFACE   xi

CHAP. 1: Introduction:
Sanguinity and Aggression   1

CHAP. 2: The NPA model   5

CHAP. 3: Genotypes and Infertility   12

CHAP. 4: Population genetics   18

CHAP. 5: Case-control studies   29

CHAP. 6: Summary   42

APPENDIX
Synopsis of NPA personality model   44

Glossary   61

References and Notes   66

Bibliography   67

Sources of Illustrations   68

Acknowledgement   69

About the Author   69

ILLUSTRATIONS & TABLES

Chapter 1: Introduction: Sanguinity and Aggression
    Fig. 1. Character types of the theory of humors

Chapter 2: The NPA model
    Fig. 2. The gingival smile of sanguinity
    Fig. 3. Faces in rage

Chapter 3: Genotypes and Infertility
    Table 1. Genotypes of Dominant types
    Table 2. Phenotypes of parents: viable progeny
    Table 3. Genotypes of non-viable types
    Table 4. Phenotypes: viable and non-viable progeny

Chapter 4: Population genetics
    Table 5. Phenotypes from gene frequencies
    Table 6. Non-viable types in two generations
    Fig. 4. NPA types in three habitancies
    Fig. 5. NPA types in four other habitancies
    Fig. 6. Successive generations in a habitancy

Chapter 5: Case-control studies
    Table 7. Gene and trait frequencies of the habitancies
    Table 8. NPA traits when Condition is related to trait A
    Table 9 & Fig. 7. Differences when Condition is related to trait A
    Table 10 & Fig. 8. …Condition is related to trait N
    Table 11 & Figs. 9-10. …Condition is related to lack of trait A or N
    Table 12. …Condition is related to trait P or lack of trait P
    Table 13. …Condition is related to an NPA type
    Table 14 & Fig. 11. …Condition has genetic heterogeneity

PREFACE

In this book, we make use of a model of personality based on genetic traits. From the title, it may seem that the book would interest only those having a technical interest in research. However, the model turns out to have some unusual features that might appeal to the general reader having an interest in human genetics. With this in mind, we have made an effort to use plain language, and we have included a synopsis of the model in the Appendix, as well as a Glossary.

First of all, the NPA model is unusual in that it is the only model of personality proffered to date that is based on classical Mendelian genetics. As we point out in the text, the conventional wisdom of the day is that there exist no genetic personality traits important enough to be worthy of a trait model of personality. Needless to say, we disagree with the conventional wisdom.

Second, the model is unusual in that two of its traits, namely sanguinity and aggression, are high frequency *recessive* traits. This leads one to the hypothesis that the two loci code for *inhibitors* of the traits.

Finally, the model is unusual in that two of its traits — again sanguinity and aggression — are genetically "entangled", in the sense that if an individual lacks one of the traits, then he or she must have the other. We show that strange things can happen in hypothetical "case-control" studies where researchers compare the genetics of a "case" group having some behavioral condition with a "control" group drawn from the general population.

The three NPA traits of sanguinity, perfectionism and aggression were first advanced as a group by Karen Horney in the 1950's. Horney's interpretation of three "expansive types" was that they were the result of stressful environment, but this is understandable, since her training was in psychoanalysis, and the age of genetics had not yet arrived. If Horney had lived to our more modern era, it is likely that she would have embraced a genetic explanation for the three traits. Very likely, she would have found great satisfaction in the idea that they are not maladaptive reactions due to onerous environment, but rather basic structural elements of the human personality.

AMB
19 December 2018

Fig. 1. Character types according to the ancient theory of humors: *Phlegmaticus, Cholericus, Sanguineus* and *Melancholicus*.
[*J.K. Lavater, ca. 1775*]

1

Introduction: Sanguinity and Aggression

PERSONALITY research has not made much progress in the past decades. There are a great many theories of personality, which means that there is no consensus at all, as to which approach is the correct one. Even the concepts of the popular theories are mostly arbitrary and empirical — and are not amenable to being either proved or falsified.

Despite controversies among those who preferentially espouse either "nature" or "nurture," it is universally acknowledged that genetics are an important facet in the etiology of the human personality. Indeed, at present, the conventional wisdom of the research community is that *many genes* contribute to personality, and the complexity is such that no gene contributes more than a few percent of the effect in any aspect of human behavior [1]. As a result, there is currently little effort being made to identify specific genetic personality traits and to incorporate them into a model of personality. The conventional wisdom has firmly established the appealing notion that the human personality is so complex that every child is effectively a *tabula rasa* at birth, capable of being molded into any conceivable configuration by parental nurture, education or other aspects of environment.

There are some lines of evidence suggesting that the above conventional wisdom is incorrect. First of all, we all know from common experience of cases where a child has a "personality type" just like one of the parents — who themselves may be very different. It is a very common occurrence. Now, if *many genes* were involved in determining basic personality type, a child's inheriting the "entire packet" of personality genes from just one of the parents would be an extremely rare event! After all, the "many" genes would be scattered on that parent's more than twenty pairs of chromosomes, and since a child inherits only one chromosome of each pair, how could "all of the many genes" be transmitted together so frequently?

Second, from the point of view of family studies: if, say, the trait of perfectionism were highly polygenic, then it would not be frequently transmitted from parent to child. But we know from experience that this is not true: very often a child diagnosed with autism or with an obsessive-compulsive personality disorder will have a parent (NP type of our model) having the same constrained personality type.

Finally, if the major personality traits were highly polygenic, then very different variations of the traits would have developed around the world, as the many genes would have assorted very differently in various populations, being subject to genetic drift and natural selection. But, that is not what one finds. Indeed, one finds that the basic personality types are the same in indigenous peoples throughout the world [2].

The NPA model: sanguinity and aggression

The NPA model is the only "trait model of personality" proposed to date that is based on classical genetics. In our experience, the model corresponds closely to reality and explains why so often a child's personality can be so similar to that of one of the parents. The model has not yet been put to test, so any results it produces must be cautiously termed as theoretical or provisional. But the potential of the NPA model is unique: it is the only theory of personality that proposes to make an assessment of the personality types of progeny based on the personality types of the parents.

The model was developed on the basis of concepts advanced over sixty years ago by psychiatrist Karen Horney. According to the model, there are three major, genetically determined, character traits that form the basis of personality. The traits are *sanguinity* (N), *perfectionism* (P) and *aggression* (A). The traits are multifaceted, each one dependent on a pleiotropic gene, by which a single gene can cause a complex pattern of physical characteristics and behavior.

The letter N is used to denote sanguinity because it is related to the classic concept of "narcissism".

The two traits, N and A, represent the fundamental basis of the human personality structure. Every individual must have a measure of trait N, or trait A, or both. Hence, the two traits are genetically *entangled*, in the sense that — in an individual or a population of individuals — lack of either trait implies the presence of the other.

The genetic basis of an individual's *NPA personality type* is determined by the combination of N, P and A traits that he or she has inherited from the parents. There are about ten common NPA personality types.

Discrete personality types

The concept that humans have a limited number of discrete character types is not new. Hippocrates in the fourth century BCE developed a system of character types based on excesses of body fluids, or "humors". The types emerged under the labels of *Sanguine, Choleric, Phlegmatic* and *Melancholic* (Fig. 1). These types are very close to the personality types generated by the NPA model, especially if one appreciates that the Sanguine and Choleric types represent the fundamental traits of sanguinity and aggression, and that the Phlegmatic and Melancholic types are composites that involve the third trait of the NPA model: perfectionism.

Entangled genes

A basic premise of the NPA model is that all individuals have an "NPA personality type" based on having a particular assortment of the three traits. However, all of the types must include either trait N or A, or both. There is no viable "pure P type"; rather the P trait appears as a modulator of the N and A traits.

The property that an individual's personality type must include either trait N or A leads to some unusual features of the model:

1. The genes for trait N and trait A are *entangled* in the sense that lack of trait N implies the presence of trait A, and vice versa.

2. Alleles corresponding to lack of expression of traits N and A act as *complementary genes,* with a *lethal effect* if both traits are not expressed in a zygote. Depending on the genotypes of the parents, some combinations of parental types can be partially infertile and some completely infertile.

3. At the level of a population, the two complementary genes could be the basis for the *speciation* of humans into two separate subspecies (Dobzhansky-Muller mechanism).

4. In *case-control studies* where factors related to personality are studied, the case and control groups are likely to have different distributions of NPA personality types, presenting difficulties in the interpretation of the results. In particular, if either sanguinity or aggression is studied, the entanglement of traits N and A could lead to serious misinterpretation of the results.

The scope of this book

Our aim is a concise presentation of the essential elements of the NPA model to show quantitatively — in terms of classical genetics — how a pair of entangled, complementary genes underlying two distinct personality traits can lead to infertility, as well as to misleading results when personality traits is studied in case-control studies.

In the chapters that follow, we present:

- The model of personality based on the three traits: N, P and A, and a brief description of the Dominant personality types generated by the model.

- Quantitative aspects of the model, namely of how the NPA traits are transmitted from parent to child and how infertility can arise.

- Population genetics: computations of distributions of NPA types, both viable and non-viable, in subpopulations on the basis of gene frequencies, with the use of Hardy-Weinberg methodology.

- Quantitative assessment of case-control studies in behavioral genetics, in particular when the object of study is a condition related to one of the entangled traits, sanguinity or aggression.

- Summary.

2

The NPA model

The NPA model of personality was developed on the basis of concepts advanced by German-American psychiatrist Karen Horney during the mid-twentieth century [3]. According to the theory, there are three major genetically determined character traits that form the basis of personality. The traits are *sanguinity* (N), *perfectionism* (P) and *aggression* (A).

An important premise of the model is that in any individual either the trait N or A, or both, must be expressed.

The three traits

Sanguinity (N) is the trait of sociability. Individuals with the trait tend to be prone to flushing, blushing and tearfulness. A hallmark of the trait is the *gingival smile* (Fig. 3) broadly exposing gums and teeth [4]. In the extreme, the trait appears as a "search for glory", and individuals may display vanity, exhibitionism and show overt narcissistic behavior. Individuals having trait N are called "sanguine" types and sometimes, appropriately, "narcissistic" types.

Aggression (A) is the well-known trait of competitiveness, often physical in nature. Individuals having the A trait (but lacking the N trait) tend to be inhibited in sociability and in flushing, blushing, tearfulness and smiling. In the extreme, the trait is a "search for power", and individuals may display physical confrontation, pugnacity and show overtly sadistic behavior. Individuals with the trait of aggression instinctively form "pecking orders". Individuals having trait A but lacking trait N are called "non-sanguine" types.

[*Max.thinks.sees*]

Fig. 2. Physiognomy of the N trait: the gingival smile. The NPA model posits that the N trait, or sanguinity, is transmitted from parents to child as a high-frequency Mendelian recessive trait.

Perfectionism (P) is a trait that may or may not be present in a given individual. It may be thought of as modulating the "unbridled" N and A traits. Individuals having overt expression of the P trait tend to value order, neatness and symmetry, and may be prone to repetitive mannerisms. In the extreme, the trait may be the cause of obsessive-compulsive or autistic-like behavior that may overwhelm other character traits. Individuals lacking trait P are called "non-perfectionistic".

Traits A and N are associated with rage reactions, namely the classic "aggressive-vindictive rage" (A rage) associated with pallor in individuals of light skin color, and the florid "narcissistic rage" (N rage) associated with sanguinity. The P trait is not associated with a rage reaction.

The traits A and N form the basis of human ambition, namely the desire to achieve power and glory, respectively.

An important result is that the model produces a limited number of discrete character types, according to how the three traits are assorted, and whether the traits are present, absent, or incompletely expressed. In this book, we shall concentrate on *Dominant types,* in which the three traits are fully expressed.

Dominant types

Dominant types are those in which all three NPA traits are either absent or fully expressed, as follows:

N   sanguine
A   non-sanguine aggressive
NA   sanguine aggressive
NP   sanguine perfectionistic
PA   non-sanguine perfectionistic aggressive
NPA   sanguine perfectionistic aggressive

Types are denoted *sanguine* or *non-sanguine* depending on the presence or absence of the trait N, respectively. The two Dominant non-sanguine types are A and PA.

Types are denoted *aggressive* or *non-aggressive* depending on the presence or absence of the trait A. The two Dominant non-aggressive types are N and NP.

Thus, there are four sanguine and two non-sanguine Dominant types, as well as four aggressive and two non-aggressive Dominant types.

Types are denoted *perfectionistic* depending on the presence or absence of the trait P. The N, A and NA types are called *non-perfectionistic* types. These three types, where neither trait N nor A is tempered by the P trait, are prone to what we term "unbridled narcissism" or "unbridled aggression".

Other types of the model

There are several other categories of character types in the NPA model. These are 1) Passive Aggressive and Resigned types, in which the A trait is partially inhibited, and 2) Borderline types, in which neither trait N nor A is fully expressed.

In order to avoid making our analyses needlessly complicated, we confine our discussion in this book to the Dominant types of the model. For further information about the other types, see the *Appendix*.

The N and A rages

The occurrence of the N and A rages is possible in any of the types having the N and A traits, respectively (Fig. 3). The rages are typically triggered by stressful environmental circumstances during an individual's daily life. The two rages can occur together, synergistically, in the "NA rage" if both the N and A traits are present in the individual's type, as for example in the NA and NPA types.

The non-aggressive N and NP types are genetically inhibited from exhibiting the A rage of aggression, just as the non-sanguine A and PA types lack the capacity to exhibit the N rage of sanguinity.

Complexities: other genes and environment

Biological variability and outliers

Just as all males and all females are not alike, there can be considerable variability in the behavioral characteristics of a given genetic

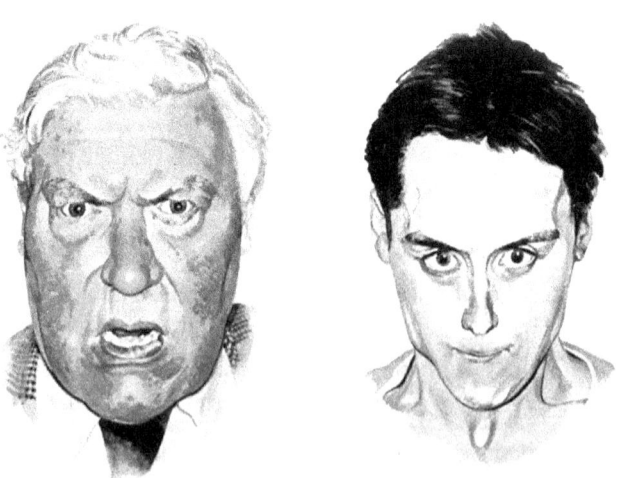

[*A. Moore*]

Fig. 3. Faces in rage: the "N rage" in a sanguine type, and "A rage" in a non-sanguine type.

NPA type. This is basically because of 1) "genes other than the NPA genes" that influence behavior, and 2) environment.

In the extreme, we should be aware that "outliers" can certainly exist. An outlier is an individual of a certain NPA type who has some unusual attributes that seem "out of character" for that particular type. There are two main reasons for outliers. First, the individual may have an unusual "other gene" that infrequently occurs in the general population, or the individual may have been exposed to an unusual set of environmental conditions. Second, the individual may simply have an unusual combination, or "perfect storm", of commonly occurring genes or environmental exposures.

Thus, the NPA traits are only a basic structural skeleton of the human personality, with many other factors, both genetic and environmental, possibly contributing to biological variability in the various NPA types. Among these are *basic drives* (hunger, thirst, sex, territoriality), *cognition* (thinking, learning, reasoning, intelligence), *temperament* (the natural activity or excitability of an individual), as well as other less clearly defined human traits, like empathy and altruism. *Environmental variables* like nurture, culture, and the individual's real-life situation in society provide a final overlay of complexity.

Temperament as a facet of personality

One of the most important aspects of the genetics of personality in the category of "genes other than the NPA genes" is the notion of temperament. By this, we mean the *general activity or reactivity* of an individual, in the sense that it is applied with regard to domesticated animals, such as dogs or horses. Thus, a particular individual, say an NP type, could be described as having a "high temperament" or a "low temperament".

The concept of temperament has not been adequately investigated in the behavioral sciences. In the NPA model we make the simplest assumption: that the genes that underlie temperament are separate from the NPA genes, hence that the NPA personality type of an individual can be determined irrespective of his or her innate level of temperament.

The "television set" analogy

One can think of an individual's personality in terms of the analogy of viewing a television receiver. If there were only two basic models of television sets, then this would represent the *male-female dichotomy*: Then, the *NPA personality type* would then be the channel selector, *temperament* would be the volume control, and how well the TV picture is actually visualized would depend on the lighting in the room, or *environment*.

While acknowledging the complexity of personality in the broader sense of the term, including the above concept of temperament, the model implies that it is the *male-female dichotomy* and the *NPA personality type* that comprise the highest genetic tiers of the human personality structure. Specifically, despite the complexities of "biological variability", temperament and other factors influencing personality, we can, in principle, identify an individual's unique genetic NPA type.

Properties of the NPA phenotypes

The pure N and A types

The N type and the A type may be regarded to be "pure" types, in the sense that they are the only types having just a single NPA trait that is not influenced by the other two traits. Thus, the traits of sanguinity and aggression can be best appreciated in the N and A types. In these typically extroverted individuals we can appreciate the unfettered extremes of human behavior that have their roots in the N and A traits. For the N trait, these extremes lie in vanity, exhibitionism and narcissism, or "narcissistic personality disorder". For the A trait, the extremes lie in coerciveness, brutishness and sadism, or "antisocial personality disorder".

The N and A traits together

The N and A traits are present together, fully expressed, in the NA type. The two traits tend not to interfere with each other, or modify each other. Rather, they appear together in an unchanged or synergistic manner, so that the NA type is typically an active, highly extroverted, non-perfectionistic individual where full-blown "unbridled" narcissism and aggression are often both on display.

The bridling effect of the P trait: NP and PA types

The effect of the presence of the P trait on the N and A traits can be profound. The effect on the N trait is such that instead of an outgoing N individual prone to vanity, the result is a less extroverted NP individual prone to perfectionistic, obsessive compulsiveness. The effect on the A trait is such that instead of an outgoing A individual prone to overt brutishness, the result is a less extroverted PA individual prone to repressed aggressive behavior.

When all three N, P and A traits are present

The individual having all three traits together, fully expressed, is the NPA dominant type (sometimes denoted as the *NPA+* type for clarity).

The resultant effect of all three traits being present together may be appreciated by imagining "adding the P trait" to the behavior of the NA type. The effect is a tempering one, but the result is still an extroverted individual who may be prone to excesses characteristic of the both the N and A traits acting in concert. The outward effects of the N and A traits may be so overt that although these individuals may consider themselves to be "perfectionists", this may not be the opinion of others. That is, the N and A traits acting together may mask the presence of the P trait as a modulating trait in the sense of "perfectionism".

Identification of NPA types

The various NPA types of the model often appear in real life in typical, stereotypical behavior that, of course, can vary according to the individual's ethnicity and real-life situation. In our previous work [5], we presented profiles of the various NPA types in the form of *caricatures*. The reason for the use of caricatures was that it allowed us to focus on the specific characteristics, and foibles, of the various NPA types without any implication that the descriptions are to be taken literally or pejoratively.

As yet, there exist no objective genetic or laboratory tests for identification of either NPA genotypes or phenotypes. The best that we can do, for the time being, is to use approximate, sometimes subjective methods [6]. In the chapters that follow, however, we set aside this important matter and focus on the theoretical predictions of the NPA model, particularly as it applies to the entangled N and A traits.

3

Genotypes and infertility

On the basis of representative family pedigrees, we posited that the NPA traits are inherited according to the mechanisms of classical genetics [7]. In particular, the three traits:

1) obey the rule of independent assortment, and

2) follow an autosomal mechanism of transmission, with traits N and A being recessive, and P being dominant.

In short, we assume that the NPA traits obey the traditional laws of Mendelian genetics, with the single genes underlying the traits being neither chromosomally linked, nor sex-linked.

The possible combinations of genes ("genotypes") consistent with the various NPA types ("phenotypes") are shown below in Table 1. The phenotypes listed are the Dominant types of the model, corresponding to full expression of the three NPA traits.

The recessive alleles of the N and A traits are denoted by **n** and **a**, while the dominant allele of the P trait is denoted by the capital letter **P**.

TABLE 1
Genotypes of the Dominant NPA types

Phenotype	*Genotype*
N	(nn) (nna)
A	(aa) (naa)
NA	(nnaa)
NP	(nnP) (nnPP) (nnPa) (nnPPa)
PA	(Paa) (PPaa) (nPaa) (nPPaa)
NPA	(nnPaa) (nnPPaa)

Table 1 shows that an individual of a particular NPA type could have one of several possible genotypes. For example, the N type has two possibilities for its underlying genotype, namely (**nn**) and (**nna**), depending on whether or not the individual is a carrier of the recessive **a** allele. In contrast, the NP type has four possibilities, depending, in addition, whether the dominant P allele is present in the homozygous or heterozygous state. The NA type has only one possible genotype, the fully homozygous (**nnaa**).

The unusual mechanism of transmission for N and A, as high-frequency *recessive* traits, leads to the hypothesis that these loci actually code for *inhibitors* of the traits, and that an inactive inhibitor would lead to a "release of inhibition" allowing expression of that trait. For trait A, the model implies that whatever the complexity of the many possible genes that permit the expression and modulation of the trait of aggression, it is a single genetic locus (the A locus of the NPA model) that permits inhibition of the final common pathway to expression of the A trait and A rage, permitting the occurrence of the non-aggressive N and NP types of the model. For trait N, the model implies that whatever the complexity of the genes that permit the expression and modulation of the trait of sanguinity, it is a single genetic locus (the N locus) that permits inhibition of the final common pathway to expression of the N trait and N rage, permitting the occurrence of the non-sanguine A and PA types.

In this book we use mainly the recessive alleles **n** and **a,** rather than their matched dominant alleles N_0 and A_0 that are assumed to code for the inhibition of traits N and A, respectively.

Offspring of Dominant NPA types

For convenience, we have tabulated the possible NPA types of the offspring according to the NPA types of the parents. Allowing for all possible genotypes in the parents, we present in Table 2 below the possible NPA types in the progeny. To read the table, one locates the NPA types of the father and mother on the horizontal and vertical axes, with the area of intersection displaying the possible NPA types in the offspring. For example, one can see that in a parental mating of NA×NP, the progeny could be only of the four types: N, NP, NA or NPA.

One can see from Table 2 that the possible NPA types of the offspring vary markedly according to the types of the parents.

Table 2 is a simplified representation of the mechanisms of inheritance of the model. Namely, it is restricted to Dominant types, hence neglects NPA types in whom the traits N or A may be incompletely expressed. However, neglecting this complexity of the NPA model does not interfere with our analyses in the chapters that follow.

	N	A	NP	NA	PA	NPA
N	N NA	"	"	"	"	"
A	N NA A	NA A	"	"	"	"
NP	N NP NA NPA	N NP NA NPA PA A	N NP NA NPA	"	"	"
NA	N NA	NA A	N NP NA NPA	NA	"	"
PA	N NP NA NPA PA A	N NP NA NPA PA A	N NP NA NPA PA A	NA NPA PA A	NA NPA PA A	"
NPA	N NP NA NPA	NA NPA PA A	N NP NA NPA	NA NPA	NA NPA PA A	NA NPA
FATHER OR MOTHER	N	A	NP	NA	PA	NPA

Table 2. Dominant types: Possible NPA types in offspring according to the types of the parents (father or mother on either axis).

Non-viable types and infertility

In addition to the six possible NPA types in the offspring shown in Table 2, the model generates two other phenotypes that are unlike any of the parental types. These are the P and 0 (null) "non-viable types", as shown in Table 3 below. We recognize these hybrid types as being "non-viable" because they lack expression of either the N or A trait.

TABLE 3
Non-viable types in progeny

Phenotype	*Genotype*
P	(nPa) (nPPa)
0 (null)	(na)

Therefore, the model quite unexpectedly predicts *infertility* in parents of certain combinations of NPA types, namely in those parental pairs who, because of their particular genotypes, are prone to conceive non-viable progeny of either the P or null phenotype, i.e., totally lacking both traits N and A. We presume that a fetus lacking expression of both of these traits would not survive intrauterine life, appearing as a miscarriage or stillbirth, or would "fail to thrive" in early infancy.

From the genotypes given in Tables 1 and 3, it can be shown that such infertility could occur only *in the mating of a non-aggressive type with a non-sanguine type*, namely N×A, N×PA, NP×A and NP×PA. Depending on the exact genotypes of the parents, infertility on this basis could be partial or complete.

In the category of *partial infertility*, one such parental combination would be a match of the genotypes **(nP/nP)×(nPa/Pa)**, which would correspond to a mating NP×PA. About fifty percent of the issue would be of genotype **(nP/Pa)**, hence non-viable. In the category of *complete infertility*, an example would be the match **(n/nP)×(a/Pa)**, which would again correspond to a mating NP×PA. Such a union could give issue only to progeny of genotype **(n/a)**, i.e., null type, or **(n/Pa)**, **(nP/a)** and **(nP/Pa)**, i.e., P types.

If we update Table 2 to include the possible non-viable P and null types in the offspring, the result is Table 4.

N	N NA	"	"	"	"	"	"
A	N NA 0 A	NA A	"	"	"	"	
NP	N NP NA NPA	N NP P NA NPA PA 0 A	N NP NA NPA	"	"	"	
NA	N NA	NA	NA A	N NP NA NPA	NA	"	"
PA	N NP P NA NPA PA 0 A	NA NPA PA A	N NP P NA NPA PA 0 A	NA NPA PA A	NA NPA PA A	"	
NPA	N NP NA NPA	NA NPA PA A	N NP NA NPA	NA NPA	NA NPA PA A	NA NPA	
FATHER OR MOTHER	N	A	NP	NA	PA	NPA	

Table 4. Dominant types: possible NPA types in offspring, including the non-viable P and 0 (null) types. Encircled are possible offspring of the four parental matches that could conceive non-viable types. These matches are N×A, N×PA, NP×A and NP×PA.

General rules pertaining to Table 4

On the basis of the NPA model, with the P trait being Mendelian dominant and the N and A traits being recessive, we can state the following general rules:

- If both parents have the N trait, then all children must also have the N trait. That is, if both parents are "sanguine", all children must be sanguine.

- If both parents have the A trait, then all children must also have the A trait. That is, if both parents are "aggressive", all children must be aggressive.

- If both parents lack the P trait, then all children must also lack the P trait. That is, if both parents are "non-perfectionistic", all children must be non-perfectionistic.

Or conversely, the rules are:

- A non-sanguine child (who lacks the N trait) must have a parent who is also non-sanguine.

- A non-aggressive child (who lacks the A trait) must have a parent who is also non-aggressive.

- A perfectionistic child (who has the P trait) must have a parent who is also perfectionistic.

With regard to the "silent" carrier state:

- If a sanguine child has a non-sanguine parent, the parent must be a carrier of the recessive allele **n**.

- If an aggressive child has a non-aggressive parent, the parent must be a carrier of the recessive allele **a**.

- If a non-sanguine child has a sanguine parent, the child must be a carrier of the recessive allele **n**.

- If a non-aggressive child has an aggressive parent, the child must be a carrier of the recessive allele **a**.

With regard to infertility:

- The alleles at the N and A loci may be thought to act as *complementary genes,* with a possible lethal effect.

- Non-viable progeny can occur only if one parent is a non-sanguine type and the other is non-aggressive type.

- Neither the NA type nor the NPA type can be the parent of a non-viable type.

Some additional considerations are:
- Parents can have children of their own NPA type, but a child's NPA type is not always the same as either parent.
- The NA type can arise in the progeny of any two parental phenotypes.
- The NA type is the only one that always breeds true: any two NA parents can have only NA offspring.
- The Table assumes no specific knowledge of the genotypes in the parents. In a specific case, where the genotypes of the parents are known, or postulated, the possibilities of the NPA types of the offspring may be more limited than those shown in Table 4.

Potential importance of entangled genes and non-viable types

In the chapters above, we have presented a model in which two major personality traits, denoted N and A, are genetically entangled by a pair of underlying complementary genes. This entanglement is the basis of possible non-viable types in progeny and of potential infertility in certain combinations of parental genotypes. But how large are these infertility effects at the level of a population? Are they quantitatively significant, or are the effects just an oddity of a model, having no practical importance. This is the question that we shall broach in the following chapters.

4

Population genetics

NPA population genetics

The NPA model, consisting of two recessive traits and one dominant trait, has its own peculiarities and subtleties, some of which we consider below. In order to simplify our analysis of the distribution of the NPA traits in subpopulations, we confine ourselves here to the most common NPA phenotypes, namely the Dominant types where all three traits are fully expressed.

Subspecies, hybrids and infertility

Since the traits N and A are posited to be Mendelian recessive, the mating of a group of pure-bred non-sanguine types with a group of pure-bred non-aggressive types would produce no viable progeny (all zygotes would lack both traits N and A). Here, "pure-bred" denotes that neither group contains individuals who are genetic carriers of the other group's recessive trait. Thus, in the context of the NPA model, non-sanguine types (A and PA) and non-aggressive types (N and NP) could be considered to be two separate subspecies. However, if either group contained carriers of the other's recessive trait, then mating between the two groups could yield viable NA and NPA+ types in the progeny. In this sense, the NA and NPA+ Dominant types can be considered to be viable hybrid types, with the P and null types being non-viable hybrid types [8].

The model's somewhat unexpected result that *Homo sapiens* could potentially be separated into two subspecies is based on the same evolutionary genetic mechanism of speciation that was first proposed by Bateson in 1909 and was later more fully developed by Dobzhansky and Muller [9]. Their analysis was based on hybrid sterility, similarly on the basis of at least one pair of interacting, complementary genes.

Geographic entanglement of traits N and A

In the NPA model the traits N and A are geographically partially entangled, in the sense that the absence of one of the traits implies the presence of the other. Thus, non-sanguine populations imply a high prevalence of aggression (A and PA types), while non-aggressive populations imply a high prevalence of sanguinity (N and NP types).

Balanced polymorphism

Why would a population be composed of several different character types? If natural selection were operative, why would not just a single character type have emerged? This is a question that population geneticists have considered in other contexts over the years and leads to the concept of *balanced polymorphism*. Geneticists often quote C.E. Ford's definition:

> "Polymorphism may be defined as the occurrence together in the same habitat of two or more discontinuous forms of a species in such proportions that the rarest cannot be maintained merely by recurrent mutation" [*10*].

An example frequently cited is that of the ABO system of antigens related to human red blood cells. In that case, the polymorphism has its basis in multiple alleles existing at a single locus. In our model of discrete character types, the polymorphism is based on the combined action of three unlinked genes.

The population dynamics relevant to the maintenance of a stable polymorphism of phenotypes within the confines of the NPA model are likely to be complex. Some relevant issues are as follows:

- In a subpopulation where certain phenotypes are at a reproductive disadvantage because of relative infertility — as by the mechanism of the NPA model — gene frequencies could nevertheless be maintained in a state of balanced polymorphism through other phenotypes by compensatory reproductive advantages, especially by assortative mating.

- Polymorphisms of character types in many areas of the world are unlikely to be in a state of balanced equilibrium because of disruptive migrations of races, nationalities and religious groups.

- Although a population may appear to be a "melting pot" of individuals in a given habitat, there could exist microcosmic isolates within the population. These isolates could have their bases in race, religion, somatotype, and especially, very directly, in the NPA personality phenotypes themselves.

- Many human societies appear to require a diversity of character types, with the various NPA types adapting with differing ease to the available stations and occupations in their societies.
- There is no apparent reason why a stable human society of a *monomorphic* character type is not possible.

One of the advantages of the NPA model is that it has at its disposal well-established quantitative techniques for the analysis populations obeying the laws of Mendelian genetics. In the following sections, we use the Hardy-Weinberg methodology to generate a variety of distributions of NPA phenotypes on the basis of assumed gene frequencies. We will use the hypothetical distributions to evaluate quantitatively how the constitution of NPA types in a subpopulation can affect the interpretation of case-control studies in behavioral genetics.

Definition of habitancy

For practical reasons we confine our analysis to the six Dominant character types (N, NP, NA, NPA, PA, A). We consider below different subpopulations where each has its own particular distribution of the various NPA phenotypes.

We define *habitancy* as the inhabitants of a region, taken collectively, i.e., a subpopulation [11]. For ease of communication we define the following habitancies:

Polymorphic — no tendency to any particular NPA type
Punctilious — trend toward NP type (sanguine)
Sublime — trend toward N type (sanguine)
Demonstrative — trend toward NPA type (sanguine)
Corybantic — trend toward NA type (sanguine)
Authoritarian — trend toward PA type (non-sanguine)
Militant — trend toward A type (non-sanguine)

We note that only the Authoritarian and Militant habitancies are appreciably non-sanguine, while only the Sublime and Punctilious habitancies are non-aggressive.

Hardy-Weinberg analysis

We used the Hardy-Weinberg approach to generate phenotype frequency distributions for the various habitancy types on the basis of assumed gene frequencies [11]. Such distributions are *idealized* balanced polymorphisms, since one assumes 1) random mating, 2) equilibrium, with allele frequencies remaining constant from one generation to the next, and 3) that the occurrence of an allele in progeny depends only on its gene frequency in the subpopulation. As applied to the NPA model, all three of these assumptions would not rigorously hold.

First of all, mating is typically highly assortative and, in fact, dependent on the personality types of individuals [12]. Second, inherent in the NPA model is the possibility of non-random infertility (Chap. 3). If the incidence of non-viable types were to be significant, NPA allele frequencies would not stay constant from one generation to the next within the confines of Hardy-Weinberg equilibrium. Third, although the NPA genes are assumed to be unlinked, certain alleles would tend to travel together according to the various NPA phenotypes in a given population. Thus, NPA alleles could appear together in a zygote less, or more, frequently than the simple prediction from their gene frequencies.

Given these complicating factors, the results of our computations of phenotype distributions are idealized. They are used primarily to give generate representative, hypothetical subpopulations having very different distributions of NPA types, and to give "order of magnitude" quantitative insight into the general nature of populations of NPA types.

With the assumption of allele frequencies n, p and a, expressions for the relative incidences of the Dominant character types are given in Table 5. In the event of the occurrence of non-viable P and null (0) phenotypes (i.e., both non-sanguine and non-aggressive types existing in the population), the condition of constant allele frequencies from one generation to the next would not be valid. In that case, the expressions in Table 5 represent the *incidences of phenotypes of the first generation only*, rather than their prevalences in a stable, balanced polymorphism.

TABLE 5
Relative Incidence of Phenotypes on Basis of Gene Frequencies

Phenotype	*Relative Incidence*
N	$n^2 \times (1-p)^2 \times (1-a^2)$
A	$(1-n^2) \times (1-p)^2 \times a^2$
NP	$n^2 \times p(2-p) \times (1-a^2)$
NA	$n^2 \times (1-p)^2 \times a^2$
PA	$(1-n^2) \times p(2-p) \times a^2$
NPA	$n^2 \times p(2-p) \times a^2$
P	$2n(1-n) \times p(2-p) \times 2a(1-a)$
Null	$2n(1-n) \times (1-p)^2 \times 2a(1-a)$

As indicated in the table, the incidence for each phenotype is the product of three probabilities, corresponding to the presence or absence of the three traits N, P and A. The P and null types contribute neither to parentage nor to viable issue.

The assumption of numerical values for the three gene frequencies generates a distribution of phenotypes for a hypothetical habitancy. Seven different habitancies were considered and given descriptive names for ease of identification: *Polymorphic, Punctilious, Sublime, Demonstrative, Corybantic, Authoritarian* and *Militant*. The intent of the labels was to emphasize the very different tenors of each of the distributions of character types. Computations for the seven habitancies are presented in Figs. 4 and 5. Each panel in the figures shows the assumed gene frequencies and the computed incidences of the NPA phenotypes (per 100 zygotes) in tabular and graphical form.

Polymorphic habitancy (mixture of Dominant types)

The assumed gene frequencies were $n = 0.90$, $p = 0.50$ and $a = 0.75$. These gene frequencies are estimates calculated from the NPA phenotype frequencies in a representative subpopulation [13]. For the Polymorphic habitancy of mixed phenotypes, the most frequent progeny types are the NPA and NP, but no type is entirely negligible. The computed frequency of the nonviable P and null types is a combined 7 percent.

Punctilious habitancy (NP type prevalent)

In relation to the Polymorphic habitancy, here we increased the value of p from 0.50 to 0.80, and decreased the value of a from 0.75 to 0.30. The incidence of the NP type is very high in this subpopulation, being about ten times greater than that of the next most frequent type, the NPA+ phenotype. The frequency of the nonviable P and null types is a combined 8 percent.

Sublime habitancy (N type prevalent)

For this habitancy we assumed values of $n = 0.99$ (highly sanguine), $p = 0.10$ (non-perfectionistic) and $a = 0.20$ (very low on aggression). The incidence of N type is high here, corresponding to over three-fourths of all progeny. Next most common is the NP type. Because of the low frequency of non-sanguinity in this subpopulation of mostly non-aggressive types, the incidence of nonviable P and null types is very low, on the order of 1 percent.

Demonstrative habitancy (NPA type prevalent)

In comparison with the Polymorphic habitancy, the assumed gene frequency of $p = 0.50$ was the same, while n was increased from 0.90 to 0.95 (more sanguine) and a was increased from 0.75 to 0.95 (more aggressive). The result is that now the NPA+ type is the most frequent in the progeny, with NA type being next most frequent. Because of the low frequencies of both non-sanguine and non-aggressive types, the incidence of nonviable P and null types is again very low, about 1 percent.

POPULATION GENETICS

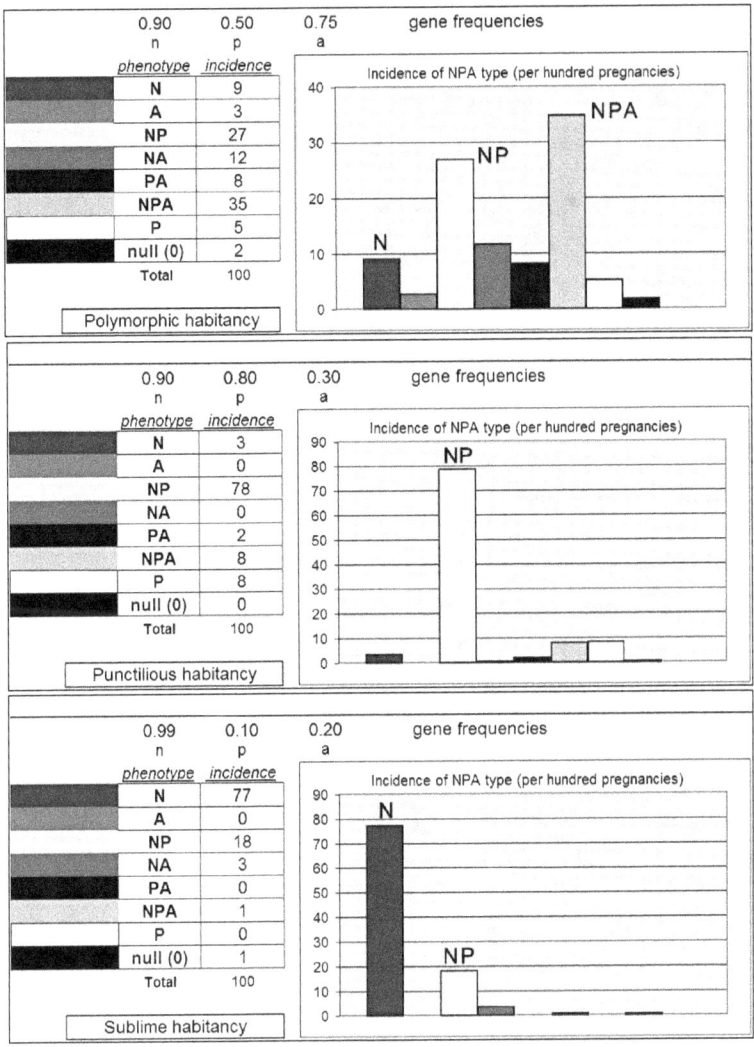

Fig. 4. Hardy-Weinberg distributions of NPA phenotypes in three hypothetical habitancies: *Polymorphic, Punctilious* and *Sublime.*

Shown at the top of each panel are the assumed gene frequencies, n, p and a (see text). Each table and bar graph shows the incidence of the NPA phenotypes, as computed from the expressions given in Table 5. The bar graphs show, from left to right, the viable types: N, A, NP, NA, PA and NPA, as well as the non-viable types: P and null. As discussed in the text, the computed Hardy-Weinberg distributions are idealized, corresponding to stable populations with random mating. We will use them in Chap. 5, as hypothetical, representative subpopulations in which case-control studies are performed.

Corybantic habitancy (NA type prevalent)

In comparison with the Demonstrative habitancy, here the assumed gene frequencies of n and a were left unchanged, while p was halved from 0.50 to 0.25 (less perfectionistic). The result was a reversal, with NA type now being the most frequent and NPA+ type being less frequent in the progeny. Once more, the computed incidence of nonviable P and null types is very low, about 1 percent.

Authoritarian habitancy (PA type prevalent)

In comparison with the Demonstrative habitancy, here we assumed gene frequencies of $a = 0.95$ and $p = 0.50$ (identical to those of the Demonstrative habitancy), but we decreased the value of n from 0.95 to 0.50 (significantly lower than that of any other habitancy). The result is that compared to the other habitancies there is a markedly higher incidence of PA and A individuals (52 and 17 percent respectively). Hence, this is predominantly a non-sanguine habitancy. The incidence of nonviable P and null types is about 5 percent.

Militant habitancy (A type prevalent)

In comparison to the Authoritarian habitancy, here we maintained the same values of gene frequencies of $n = 0.50$ and $a = 0.95$, but we decreased the value p from 0.50 to 0.20 (highly non-perfectionistic). The result was that compared to the Authoritarian habitancy — although the degree of non-sanguinity was the same (the same total of PA plus A types in the progeny) — there was a reversal, so now A types outnumber PA types (44 and 25 percent, respectively). This can be interpreted to be a marked shift from perfectionistic aggression to unbridled aggression. The incidence of nonviable P and null types remained the same, at about 5 percent.

Frequency of non-viable types in the Hardy-Weinberg habitancies

For each of the habitancies presented in Figs. 4-5, the computed frequency of non-viability was found to be in the range of 0 to 9 percent (summarized in Table 6, overleaf), i.e., negligible in some habitancies but moderately high in others. Also shown in Table 6 is the frequency of non-viable types in the subsequent generation, using gene frequencies of the viable types of its preceding generation. It is clear that these habitancies are not in equilibrium, as the frequency of non-viable types may decrease appreciably in a single generation. Although the computations indicate that in most habitancies the incidence of non-viable types would be on the order of 5 per cent or less, they raise the possibility that fetal loss due to this mechanism could indeed be significant in some areas of the world [14].

POPULATION GENETICS

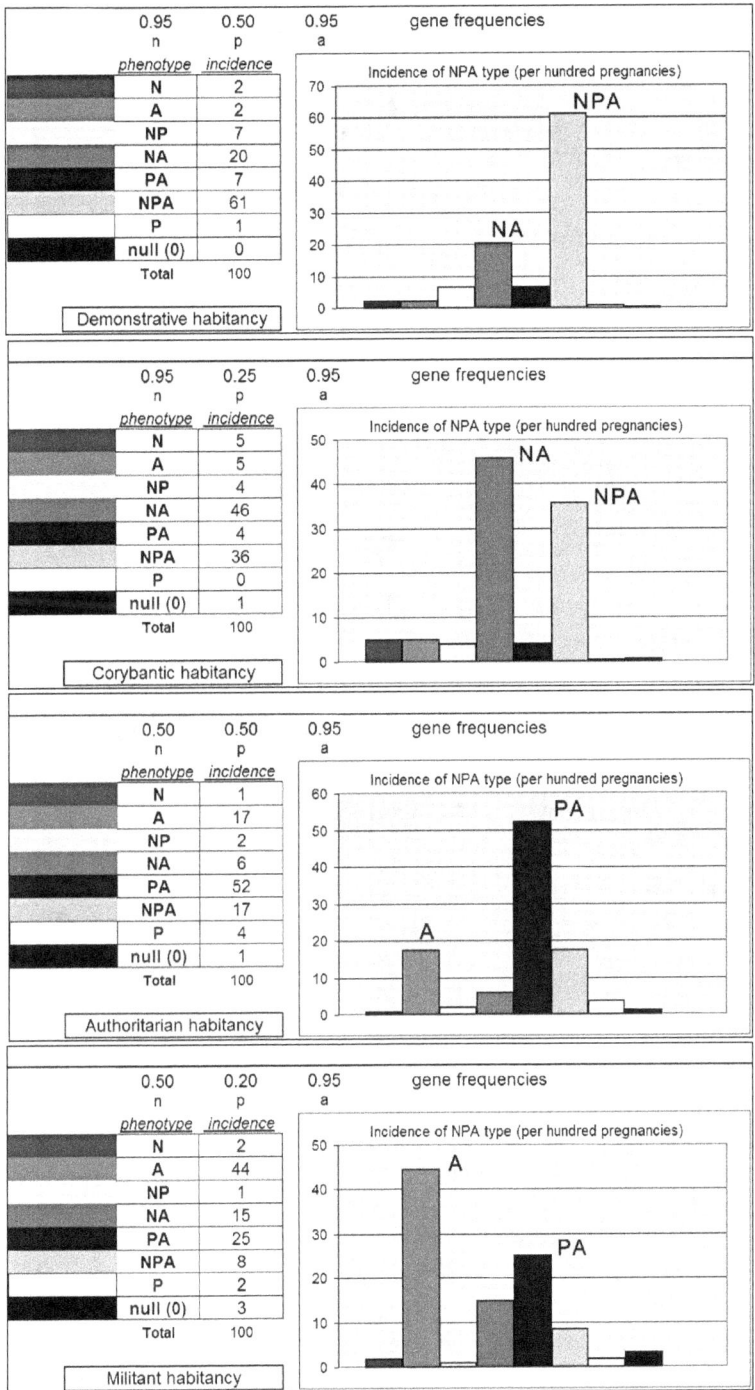

Fig. 5. Hardy-Weinberg distributions of NPA phenotypes in four additional hypothetical habitancies: *Demonstrative*, *Corybantic*, *Authoritarian* and *Militant*.

TABLE 6
Incidence of non-viable types in habitancies for two successive generations
(per 100 zygotes)

	present generation	*next generation*
Polymorphic	6.9	3.9
Punctilious	8.4	0.9
Sublime	0.6	0.03
Demonstrative	0.9	0.8
Corybantic	0.9	0.8
Authoritarian	4.9	1.3
Militant	4.9	1.3

In a habitancy that contains both non-aggressive and non-sanguine types, one can see that, with random mating, each successive generation would show a preferential increase in the frequencies of the viable hybrid NA and NPA+ types at the expense of fetal loss of the non-viable P and null zygotes. To investigate this further, we considered the extreme case of a habitancy *initially composed of only (non-aggressive) NP types and (non-sanguine) A types in equal proportions.* We conducted a Hardy-Weinberg stepwise simulation for three successive generations of progeny, taking into account the fetal loss of non-viable P and null types with each generation.

The results are shown in Fig. 6 on the next page. The bar graph in the top panel shows the initial equal prevalence of the parental NP and A types. Each successive lower panel of the figure depicts the next generation of progeny, displaying in tabular form and in bar graphs the incidence of the various NPA phenotypes, together with the changing gene frequencies.

The results for the first generation of progeny show an immediate dramatic change from an initial population composed of only NP and A types to one where all of the viable types are present in equal proportions. The incidence of non-viable P and null types is seen to be 18 percent, or almost one in five zygotes. In the following two generations there is a rapid rise in the frequency of the hybrid NA and NPA+ types, with a concomitant decrease in the non-sanguine and non-aggressive viable types, as well as a progressive decrease in the incidence of the non-viable P and null types.

The above example considers a habitancy where changes in gene frequencies occur rapidly. However, the computations suggest that in some *stable* subpopulations, where stable gene frequencies are maintained

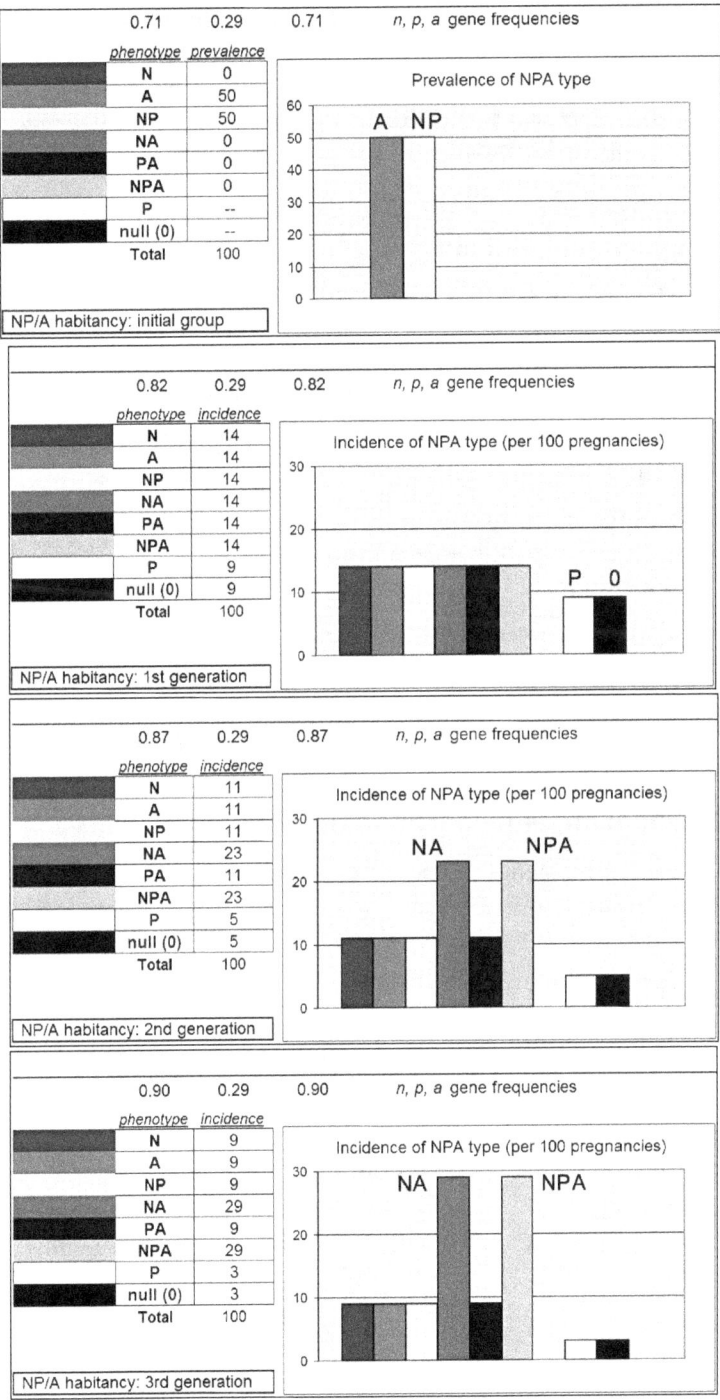

Fig. 6. Phenotype frequency distributions in successive generations for a habitancy initially of NP and A types in equal proportion. With each generation, the frequency of hybrid NA and NPA types increases.

by compensatory mechanisms, the mode of infertility inherent in the NPA model could affect the equilibrium levels of the NPA gene frequencies and phenotypes. As illustrated by the example in Fig. 6, the effect of such infertility in a subpopulation would be to *favor the trait of aggression over non-aggression*, and *sanguinity over non-sanguinity*.

Summary of Hardy-Weinberg analysis

Although the Hardy-Weinberg approach is based on idealized assumptions, it does provide some quantitative insight into the dynamics that would exist in habitancies consisting of the various distributions of NPA phenotypes.

First, the computations indicate that even modest changes in the assumed NPA gene frequencies can engender significant changes in the tenor of a habitancy.

Second, the computations suggest that infertility inherent in the NPA model, resulting in predictable non-viable phenotypes in the progeny, could be quantitatively significant in some habitancies.

Finally, the analysis indicates that in the case of significant non-viable progeny occurring as an initial condition in a habitancy, changes in gene frequencies (and phenotypes) could occur with great rapidity from generation to generation, with the NA and NPA+ hybrid types gaining a reproductive advantage.

Research studies of behavioral traits

In general, the genetic bases of behavioral traits can be investigated in 1) *family studies*, where affected and unaffected members of the family are compared, and 2) *case-control studies,* in which the control subjects are often a random sample from the general population. As family studies are relatively onerous to perform, researchers commonly rely on case-control studies. Since such studies often require a sample of control subjects drawn from the population at large, the distribution of NPA types in these control samples will be highly dependent on the habitancy in which the study is performed. This fact, together with the entanglement of the traits N and A inherent in the NPA model, gives us a guarantee that the interpretation of case-control behavioral studies will not be straightforward. In the following chapter, we address this issue quantitatively.

5

Case-control studies

The case-control study

In a typical case-control study in behavioral genetics, there are two groups of subjects: the *case* group, which has some uncommon genetic characteristic, and the *control* group, which is often drawn from the general population. The researchers use statistical techniques to compare the frequencies of suspected genetic markers in the two groups.

In this chapter we will consider specific examples of case-control studies, with a special interest to see what occurs when the case group has a "Condition" related to one of the NPA traits [15]. One can see immediately two possible sources of confusion:

- Although the case group of subjects might be similar, irrespective of where the study is conducted, the control group will have a very different distribution of NPA types depending on the habitancy where the study is conducted.
- If the Condition of the case group is related to trait N or to trait A, then entanglement between the two traits could lead to misinterpretation of the results.

In the examples that follow, we assume 1) that the Condition, such as a disease or behavioral disorder, is related to some aspect of the NPA traits, and 2) we assume that all subjects of the control groups are drawn from random samples of the general population of the habitancy in which the study is conducted.

Although geneticists usually compare the *genotypes* (allele frequencies) of the case and control groups, we shall focus on differences in NPA *phenotypes*. This simplifies the computations without affecting any of our conclusions.

Examples of studies

We assume that the researchers can measure precisely the genotypes of the N, P and A traits in their subjects of both the case and control groups, but they have no idea to which trait (if any) the Condition is related.

Below, we consider six idealized studies where the Condition is actually related 1) to traits N or A, or to their absence, 2) to trait P or its absence, and 3) to a particular NPA type.

Study 1: Condition related to trait A

Researchers conduct a study where, unbeknownst to them, the Condition (a disorder of behavior) is causally related to trait A, such that any individual in the habitancy who has trait A has the same risk of having the Condition. We assume that the identification of the Condition is a very clear-cut diagnosis. The researchers compare the frequencies of all three NPA traits (actually genotypes) between the study and control group.

To begin, we present in Table 7 a summary of the gene frequencies used in generating the Habitancies, as well as the NPA trait frequencies for each habitancy. For example, for the Polymorphic habitancy, 88 of every hundred individuals have the N trait, and so on. We note that the trait frequencies vary widely according to habitancy. For example, the frequency of A trait is 97% in the Authoritarian habitancy but only 4% in the Sublime habitancy. Since our control groups are drawn from the general population, *the trait frequencies in Table 7 are those of the control groups of our hypothetical studies*.

TABLE 7
NPA gene frequencies and trait frequencies for viable types in the various habitancies

habitancy	*gene frequency*			*trait frequency (%)*		
	n	*p*	*a*	N	P	A
Polymorphic	0.90	0.50	0.75	88	75	61
Punctilious	0.90	0.80	0.30	98	96	11
Sublime	0.99	0.10	0.20	99	19	4
Demonstrative	0.95	0.50	0.95	91	75	91
Corybantic	0.95	0.25	0.95	91	44	91
Authoritarian	0.50	0.50	0.95	27	75	97
Militant	0.50	0.20	0.95	27	36	97

Next, we present Table 8, showing the NPA trait frequencies for the case and control groups of Study 1. As the Condition is related to trait A, Table 8 indicates that 100% of the individuals in the case group have trait A, irrespective of habitancy.

Finally, using the data of Table 8, we present in Table 9 the *difference* (delta) in the NPA traits between the case and control groups. The results in Table 9 are also presented graphically in Fig. 7, overleaf.

We note the following:

TABLE 8

Frequency of traits N, P and A in case and control groups when Condition is related to trait A in various habitancies (percent)

habitancy	case			control		
	N	P	A	N	P	A
Polymorphic	81.0	75.0	100	88.3	75.0	61.3
Punctilious	81.0	96.0	100	97.9	96.0	10.9
Sublime	81.0	19.0	100	99.9	19.0	4.1
Demonstrative	90.3	75.0	100	91.1	75.0	91.1
Corybantic	90.3	43.8	100	91.1	43.8	91.1
Authoritarian	25.0	75.0	100	27.0	75.0	97.4
Militant	25.0	36.0	100	27.0	36.0	97.4

TABLE 9

Difference (delta) in frequency of traits N, P and A between case and control groups when Condition is related to trait A in various habitancies

habitancy	N	P	A
Polymorphic	-7.3	0	38.7
Punctilious	-16.9	0	89.1
Sublime	-18.9	0	95.9
Demonstrative	-0.9	0	8.9
Corybantic	-0.9	0	8.9
Authoritarian	-2.0	0	2.6
Militant	-2.0	0	2.6

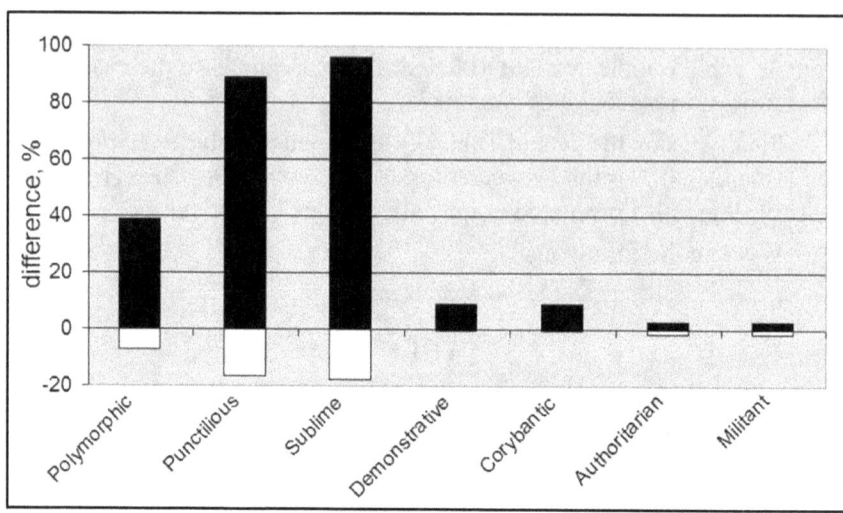

Fig. 7. Difference in frequency of traits A and N between case and control groups when Condition is related to *trait A* in various habitancies. The black upward bars represent the differences in trait A for the various habitancies. The white downward bars represent differences in trait N, which is a reciprocal effect due solely to the entanglement of trait A with trait N.

- For trait A, the values of *delta A* are variable, being large for some habitancies, but low in others. The values are highest for the Punctilious and Sublime habitancies, where the control group has a very low prevalence of trait A, and lowest in the Authoritarian and Militant habitancies, where the control group has a high prevalence of trait A. Thus, depending on the habitancy where the study was performed, the researchers could draw very different conclusions as to whether the Condition was related, or not related, to trait A.

- Despite the fact that the Condition is genetically linked only to trait A, Table 9 and Fig. 7 show that there are sizable differences in *delta N*, as well. The values of delta N are most marked for the Punctilious and Sublime habitancies, where, because of the entanglement of traits N and A, the study group has a lower frequency of trait N. If the researchers had measured trait N, but not trait A, they could have mistakenly concluded that the Condition was related to the N locus, rather than the A locus — but only in some of the habitancies in which the study was performed.

- For trait P, the values of *delta P* are zero for all of the habitancies. This follows from the fact that trait P is not entangled with either N or A, with the frequency of the P trait being identical in the study group and in the general population. Thus, if the researchers had measured trait P, irrespective of the habitancy where the study was performed, they would have correctly concluded that there was no evidence that the Condition was related to trait P.

To summarize: when the Condition is genetically related to trait A: 1) the results of the study are highly dependent on the habitancy in which the study is performed, 2) because of entanglement of traits N and A, the researchers may mistakenly conclude that the condition is related to the gene for trait N at a very different locus, 3) despite being genetically involved as an essential part of NPA typology, trait P is not entangled with traits N or A, and it does not show up as a difference between the case and control groups, and 4) a Condition related to trait A is best studied in a habitancy where the prevalence of the trait is low.

Study 2: Condition related to trait N

Similar to Study 1, researchers conduct a study where, unbeknownst to them, the Condition is actually related to trait N, such that any individual in the habitancy who has trait N has the same likelihood of having the Condition. Again, we assume that the identification of the Condition is a very clear-cut diagnosis. The researchers measure the frequencies of all the traits N, P and A in both the study and control groups.

As in the previous study, we present in Table 10 and in Fig. 8, overleaf, the difference (delta) in the NPA trait frequencies between the case and control groups. Here, we find again that the delta for trait N is very large for some habitancies but very small in others, depending on whether the trait has a low or high prevalence in the general population. Again, we find that the entangled trait A shows significant reciprocal values of delta A. The values, however, are more modest than in Study 1, being only as high as 5 to 7 percent. Again, the unentangled trait P shows delta values of zero for all habitancies.

Thus, when the Condition is genetically related to trait N, we conclude that: 1) the results of the study are highly dependent on the habitancy in which the study is performed, 2) because of entanglement of traits N and A, the researchers may mistakenly conclude that the condition is somewhat related to the gene for trait A, 3) trait P shows no entanglement with traits N or A, and 4) a Condition related to trait N is best studied in a non-sanguine habitancy, where the prevalence of the trait is low.

TABLE 10

Difference (delta) in frequency of traits N, P and A between case and control groups when Condition is related to trait N in various habitancies

habitancy	N	P	A
Polymorphic	11.7	0	-5.1
Punctilious	2.1	0	-1.9
Sublime	0.1	0	-0.1
Demonstrative	8.9	0	-0.9
Corybantic	8.9	0	-0.9
Authoritarian	73.0	0	-7.1
Militant	73.0	0	-7.1

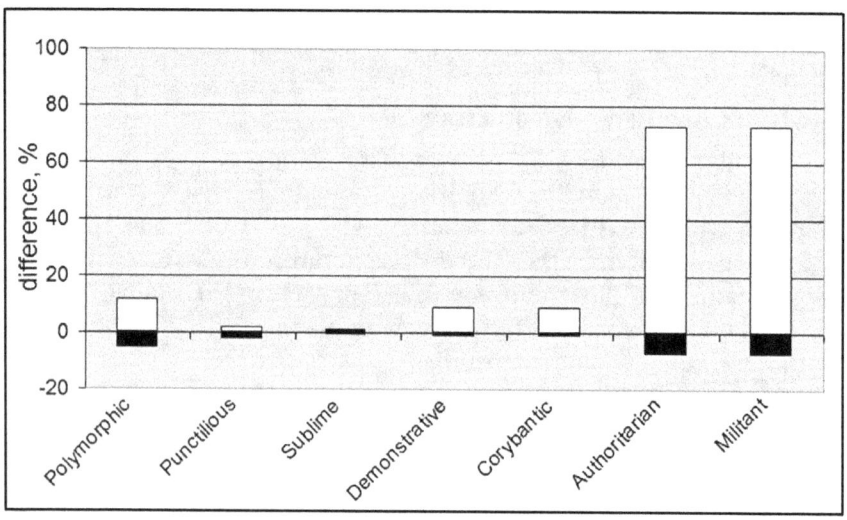

Fig. 8. Difference in frequency of traits N and A between case and control groups when Condition is related to *trait N* in various habitancies. The white upward bars represent the differences in trait N. The black downward bars represent reciprocal differences in trait A due solely to the entanglement of trait N with trait A.

Study 3: Conditions related to the lack of one of the traits, A or N

These studies would be similar to Studies 1 and 2, except now the Condition is related to the *absence* of one of the traits, A or N.

Lack of trait A implies that the case group is limited to the non-aggressive N and NP types, whereas lack of trait N in another study implies that the case group is limited to the non-sanguine A and PA types.

The relevant computations for the NPA frequency differences between the case and control groups are shown in Table 11 and in Figs. 9-10, overleaf.

TABLE 11
Difference (delta) in frequency of traits N, P and A between case and control groups when Condition is related to lack of trait A or N in various habitancies

habitancy	lack of trait A			lack of trait N		
	N	P	A	N	P	A
Polymorphic	11.7	0	-61.3	-88.3	0	38.7
Punctilious	2.1	0	-10.9	-97.9	0	89.1
Sublime	0.1	0	-4.1	-99.0	0	95.9
Demonstrative	8.9	0	-91.1	-91.1	0	8.9
Corybantic	8.9	0	-91.1	-91.1	0	8.9
Authoritarian	73.0	0	-97.4	-27.0	0	2.6
Militant	73.0	0	-97.4	-27.0	0	2.6

We note the following:

- As in Studies 1 and 2, the delta N and A values are highly dependent on the habitancy in which the study is performed.
- Not surprisingly, the entanglement between traits N and A is found to be even greater when the Condition is related to the *absence* of either trait, as all of the subjects in the case groups would have only one of the traits, either A or N. For the Authoritarian habitancy, *lack of trait A* in the case group shows up as a large delta N difference of 73 percent due to entanglement, while for the Sublime habitancy, *lack of trait N* in the case group shows up as a delta A difference as large as 96 percent. In these habitancies, if the researchers measured only one of the traits, N or A, they could easily misconstrue the results, concluding that the Condition was related to the entangled trait.
- As in Studies 1 and 2, trait P is not entangled with either trait N or A, and the delta P values are nil for all habitancies.

Thus, when the Condition is related to *absence* of either trait A or N, the same issues are encountered as in Studies 1 and 2, with the added comment that the quantitative entanglement between the traits can be even stronger.

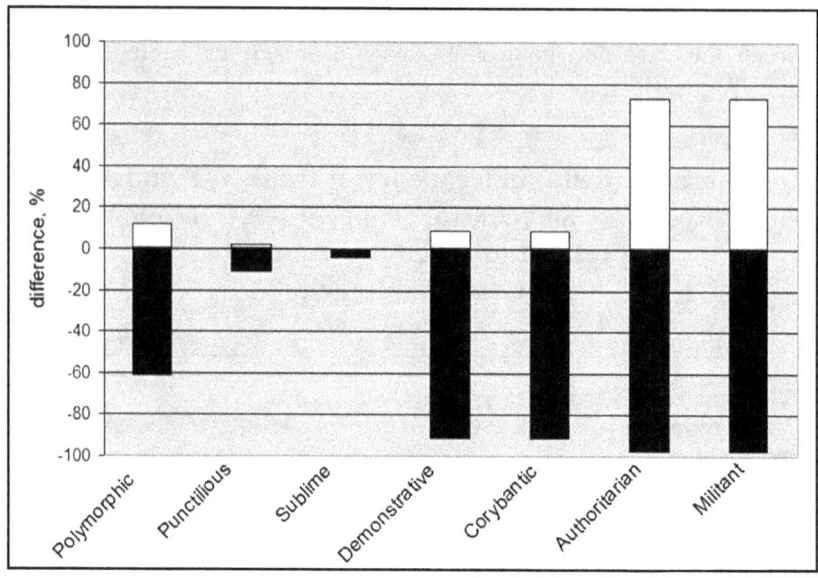

Fig. 9. Difference in frequency of traits N and A between case and control groups when Condition is related to the *absence of trait A*. The black downward bars represent the differences in trait A. The white upward bars represent reciprocal differences in trait N due solely to the entanglement of trait A with trait N.

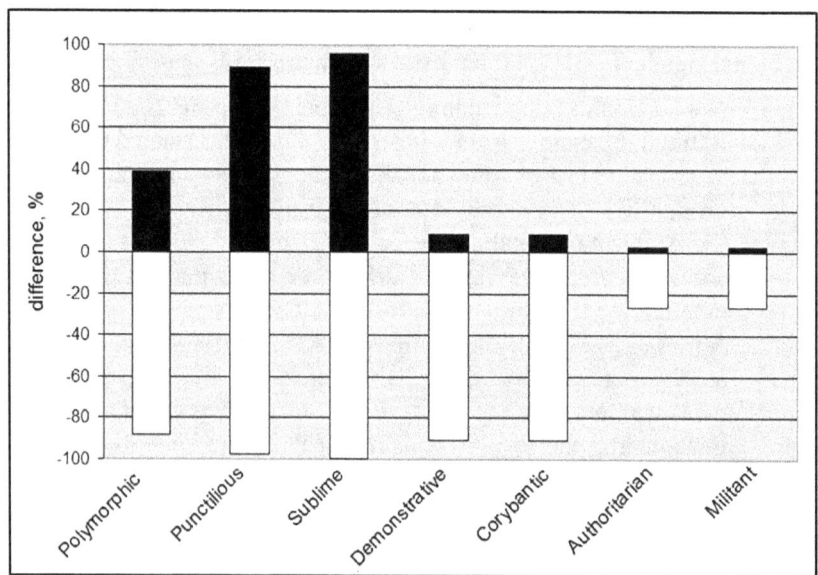

Fig. 10. Difference in frequency of traits N and A between case and control groups when Condition is related to the *absence of trait N*. The white downward bars represent the differences in trait N. The black upward bars represent reciprocal differences in trait A due solely to entanglement.

Study 4: Condition related to trait P or absence of trait P

This study would be similar to Studies 1-3, except now the Condition is related to the presence or absence of the trait P.

In Studies 1-3, trait P was found to be an unentangled "hitchhiker", assorting itself passively between the case and control groups, so that no difference in P trait showed up for any habitancy in any of the studies. Here, we consider the case where the Condition is actually genetically related to trait P or its absence. An example of a Condition related to trait P might be an obsessive-compulsive personality disorder or an eating disorder. An example of a Condition related to lack of trait P might be an attention deficit disorder that is limited to N, A and NA types.

The relevant computations for the NPA frequency differences between the case and control groups are shown in Table 12.

TABLE 12
Difference (delta) in frequency of traits N, P and A between case and control groups when Condition is related to trait P or lack of trait P

habitancy	trait P			lack of trait P		
	N	P	A	N	P	A
Polymorphic	0	25.0	0	0	-75.0	0
Punctilious	0	4.0	0	0	-96.0	0
Sublime	0	81.0	0	0	-19.0	0
Demonstrative	0	25.0	0	0	-75.0	0
Corybantic	0	56.3	0	0	-43.8	0
Authoritarian	0	25.0	0	0	-75.0	0
Militant	0	64.0	0	0	-36.0	0

The table shows that:
- The delta P values are large for most of the habitancies. When the Condition is related to P, the highest values of delta P occur for those habitancies where the prevalence of P is low. When the Condition is related to the absence of P, the highest values of delta P occur for those habitancies where the prevalence of P is high.
- There is no entanglement of trait P with traits N or A, the delta N and delta A values being zero for all habitancies.

Thus, despite being a key trait in the determination of NPA type, the P trait is partitioned unequally between the case and control groups only when the Condition is genetically related to the P locus.

CHAPTER FIVE

Study 5: Condition related to a particular NPA type

This study would be similar to Studies 1-4 above, except now the Condition is related to a particular NPA type (rather than to a trait). We consider here the example where the Condition occurs in only the *NP type,* i.e., in a type having only one of the traits, N or A.

Examples of conditions that may be linked to the NP type are a disorder in the autism spectrum, or the occurrence of a very tall child in parents of short stature [16].

The relevant computations for the NPA frequency differences between the case and control groups are shown in Table 13:

TABLE 13
Difference (delta) in frequency of traits N, P and A between case and control groups when Condition is related to NP type in various habitancies

habitancy	N	P	A
Polymorphic	11.7	25.0	-61.3
Punctilious	2.1	4.0	-10.9
Sublime	0.1	81.0	-4.1
Demonstrative	8.9	25.0	-91.1
Corybantic	8.9	56.3	-91.1
Authoritarian	73.0	25.0	-97.4
Militant	73.0	64.0	-97.4

The Table shows:

- As in the previous studies, there is a wide disparity in the results for the delta values according to the habitancy in which the study is performed.

- All three NPA traits show differences between the case and control groups. Notably, both the delta N and delta A values are *exactly the same* as in Study 3 (lack of A trait, Table 11), while the delta P values are the *exactly the same* as in Study 4 (P trait, Table 12). This emphasizes an important point: although it may appear that the NP type requires the specification of three genetic loci, it suffices to specify only two independent conditions for individuals in the NP case group: 1) lack of A trait, and 2) presence of P trait. The third requirement, presence of N trait is not independent, but follows from the entanglement of traits A and N.

Study 6: Condition related to the lack of either of two traits, N or A

This study would be similar to Study 3, except now the subjects in the case group would lack *either* of the two traits, N or A. That is, a case group could be a mix of N, NP, A and PA types.

An example of a condition that could occur preferentially in types having only one of the traits, N or A, is the mental illness, schizophrenia. In prior work [17], we posited that having only trait N, or only trait A, could augment an individual's predisposition to the genetic and environmental risks of schizophrenia. In contrast, the NA and NPA+ types, with full expression of both traits, would have a much lower predisposition.

This Study is different from the ones above, in that it is an example of *genetic heterogeneity*. That is, the Condition can be the expression of either of two separate genes.

The relevant computations for the NPA frequency differences between the case and control groups are shown in Table 14 and Fig. 11 overleaf:

We note the following:

- As in the previous studies, there is a wide disparity in the results for the delta N and delta A values according to the habitancy in which the study is performed.
- There are sizable delta values for both trait N and A between the case and control groups. However, the differences are all in the negative direction, with no reciprocal effects of entanglement. Although the two traits remain theoretically entangled by their complementary genes, the main effect here is the *causative involvement of both traits* in the case groups. In fact, the large negative delta A and N values in Table 14 and Fig. 11 are due entirely to the fact that the control groups contain NA and NPA+ types, i.e., types having both traits.
- Although this Study does not exhibit reciprocal effects of the entanglement of traits N and A, there could be problems in the interpretation of the results, nonetheless. Researchers could easily find large differences in delta N and A, and erroneously conclude that the Condition is (negatively) related to the two traits in a very straightforward manner (i.e., lower percent of both traits N and A in the case subjects). However, it would require careful scrutiny of the genotypes in both the case and control groups to correctly conclude that 1) the case subjects were a heterogeneous mix

TABLE 14

Frequency differences in NPA traits and types between case and control group when a Condition is related to lack of either trait N or A in various habitancies (percent)

habitancy	N	P	A	NA & NPA types *
Polymorphic	-11.5	0	-38.2	-46.3
Punctilious	-0.2	0	-8.6	-8.1
Sublime	-0.0	0	-4.0	-4.0
Demonstrative	-41.1	0	-41.1	-81.5
Corybantic	-41.1	0	-41.1	-81.5
Authoritarian	-23.5	0	-0.9	-23.2
Militant	-23.5	0	-0.9	-23.2

* Since the case groups contain no NA or NPA types, these numeric values represent the percent sum of NA and NPA types in the control groups.

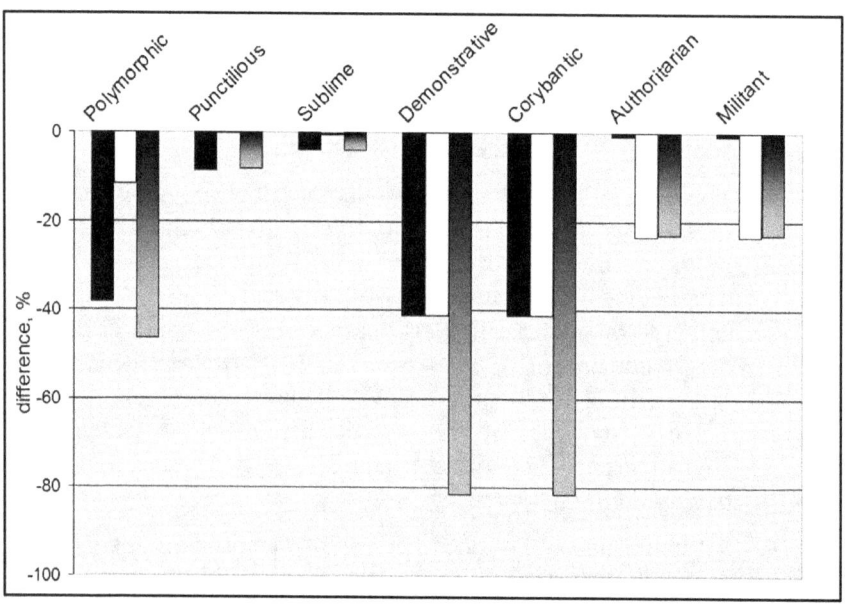

Fig. 11. Differences between a case and control group when a Condition is related to the *lack of either of the two traits, N or A*. The black and white bars represent the differences in traits A and N, respectively. The shaded bars represent the *sum of the frequencies of NA and NPA types* in the control groups. Since the NA and NPA types have low vulnerability to the Condition, the magnitudes of the shaded bars represent the percentage of the general population having genetic "protection" from the condition.

of individuals having only one of two unlinked traits, and 2) the sizable negative delta N and A values observed were solely result of the control groups' containing "protected" NA and NPA individuals having both traits.
- As in the prior Studies 1-3, trait P is not entangled with either trait N or A, the delta P values all being zero.
- The most sensitive case-control studies would be those performed in habitancies having high prevalence of the NA and NPA types, i.e., the types having both traits.
- The sum of the frequencies of the NA and NPA types in the control group represents the percentage of individuals in the habitancy who have "protection" from the Condition.
- If the Condition were to be applicable to schizophrenia in the real world, then 1) the N and A loci, acting as complementary genes, would define an N/A phenotype (both traits expressed) that represents relative protection from schizophrenia, and 2) the sum of the NA and NPA frequencies in the population at large would represent the percentage of individuals who have a measure of genetic protection from the condition.

Interpretation of results of case-control studies

The main points that become clear from our quantitative assessments of case-control studies are:

1. The results of studies may be highly dependent on the habitancy in which the study is performed.
2. When the Condition studied is related to the personality traits of sanguinity or aggression, special problems of interpretation may arise, as the two traits are entangled.

The case examples presented above are highly idealized, in the sense that the computations were done on the basis of precisely specified habitancies, with all of the control groups being random samples of the hypothetical populations. In the real world, where the case group may be genetically heterogeneous, and the control group may be a skewed sample of the population, problems of interpretation are likely to be even more severe than those that we outlined above.

From the point of view of the NPA model, any study in behavioral genetics could be "contaminated" by unrecognized distributions of NPA types in either the case or control groups, possibly inducing unwelcome associations with other genes influencing behavior.

6

Summary

- Our objective is to examine how interacting genetic traits can affect the interpretation of the results of case-control studies.

- The basis of our analysis is the NPA model of personality comprising three genetic traits: sanguinity (N), perfectionism (P) and aggression (A).

- The model is based on classical Mendelian genetics, with traits N and A being high frequency recessive, and P being dominant.

- The genetic loci for traits N and A code for *inhibitors* of the traits.

- The model produces a limited number of discrete personality phenotypes. In a personality type called a Dominant type, the NPA traits are fully expressed.

- An NPA type must contain either trait N or trait A. Thus, the genes for traits N and A are *entangled* in the sense that if one of the traits is absent, then the other must be present.

- The P trait acts as a modifier of the N and A traits; it may or may not be present in an NPA type.

- An NPA type is called *non-sanguine* if it lacks the N trait, and *non-aggressive* if it lacks the A trait.

- The genes associated with the N and A loci act as *complementary genes* having a lethal effect when neither trait is expressed in a zygote.

- Depending on their genotypes, certain parental combinations of NPA types are prone to infertility. Infertility occurs when a zygote lacks both traits N and A, hence is non-viable.

- Infertility occurs only when one parent is a non-aggressive type and the other is a non-sanguine type.

SUMMARY

- At the level of population genetics, the mechanism of infertility inherent in the NPA model is identical to the Dobzhansky-Muller model of speciation based on complementary genes. By this mechanism, the human species could theoretically be divided into two separate subspecies: one non-aggressive and the other non-sanguine.

- We define *habitancy* as a subpopulation having a certain distribution of NPA types. We used Hardy-Weinberg methodology to compute the frequencies of phenotypes in various habitancies with the use of gene frequencies.

- In the hypothetical habitancies, the computed frequency of non-viable types was low to moderate, on the order of 1 to 9 percent.

- The various hypothetical habitancies were used as the basis for examining examples of case-control studies where the studied condition was related to one of the NPA traits.

- The results of the hypothetical case-control studies were highly dependent on the habitancy type.

- When the condition characterizing the case group was genetically related to trait A, there was nevertheless a reciprocal difference induced in trait N due to the entanglement of the two traits, leading to possible confusion in interpretation of the results. Similarly, if the condition was related to trait N, there was similar reciprocal entanglement with trait A.

- When the condition studied in the case group was genetically related to the *absence of one of the traits*, A or N, the effects of entanglement were even greater.

- When the condition studied in the case group was genetically related to the *absence of either of the two traits*, A or N, there were no reciprocal effects of entanglement.

- When the condition studied in the case group was genetically related to trait P or its absence, there were no entanglement or association effects in any of the hypothetical studies, despite the trait's being an essential element of NPA typology.

APPENDIX A
NPA PERSONALITY THEORY: SYNOPSIS
Personality theory based on the genetic traits of sanguinity, perfectionism and aggression

The NPA theory of personality was developed by A.M. Benis on the basis of concepts presented over sixty years ago by psychiatrist Karen Horney. The model posits three major behavioral traits underlying personality: sanguinity (N), perfectionism (P) and aggression (A), leading to the formulation of discrete character types. Each trait is based on a major pleiotropic gene (a gene determining several related characteristics) that follows the rules of Mendelian genetics.

The NPA model proposes that the character traits A and N are indispensable to human development, being related to the sympathetic and parasympathetic nervous systems, respectively. The trait P is also assumed to function at the level of the central nervous system and to act as a modifier of the expression of traits A and N. The NPA model proposes to clarify the genetic bases of known personality disorders, diseases related to behavioral factors ("psychosomatic diseases") and mental illnesses. An online NPA personality test is available in English and other language versions.

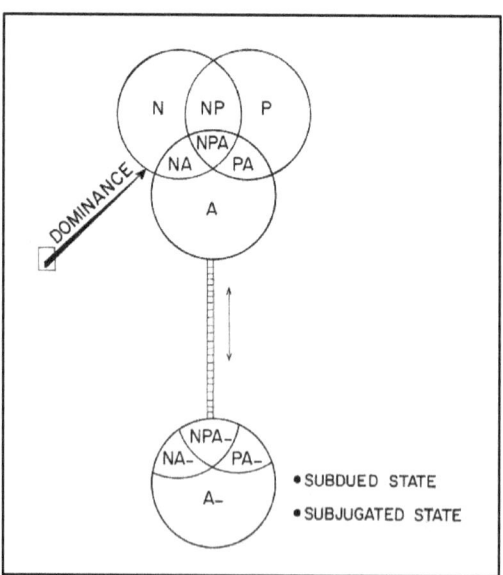

Fig. A1. Venn diagram of dominant character types. Character types having the trait of aggression A may be reduced, reversibly, to a subdued or subjugated state A–.

Contents

- 1 What is personality?
- 2 NPA model based on three genetic traits
  - 2.1 Genetics and environment
  - 2.2 Traits of sanguinity, perfectionism and aggression
    - 2.2.1 Aggression (A)
    - 2.2.2 Sanguinity (N)
    - 2.2.3 Perfectionism (P)
- 3 Character types
  - 3.1 Dominance: Dominant types
    - 3.1.1 N type
    - 3.1.2 A type
    - 3.1.3 NA type
    - 3.1.4 NP type
    - 3.1.5 PA type
    - 3.1.6 NPA type
  - 3.2 Submission: Passive Aggressive types
    - 3.2.1 Non-compliant types
    - 3.2.2 Compliant types
  - 3.3 Resignation: Resigned types
- 4 Borderline types and mental illness
- 5 Dominance and submission
- 6 Mendelian transmission of NPA traits
- 7 Implications of a trait theory based on genetics
  - 7.1 Population Genetics
  - 7.2 Evolutionary origins of NPA traits
  - 7.3 Predictive aspects of NPA model
- 8 Criticism and controversy
- 9 References
- 10 Citations
- 11 Illustrations
- 12 Source

What is personality?

Psychologists speak of personality as "a collection of emotional, thought and behavioral patterns unique to a person that is consistent over time" [1]. Although many investigators have proposed various theories of personality, no objectively testable model has emerged. The NPA model falls into the category of a trait theory of personality, its unique approach being that it is biologically based on classical human genetics.

NPA model based on three genetic traits

Genetics and environment

Although it is universally accepted that both genetic and environmental factors (or "nature and nurture") comprise personality, the relevant genes have yet to be identified [2]. Studies of the heritability of personality factors conducted with identical and fraternal twins emphasize the importance of genetics in behavior [3]. The NPA model acknowledges the possible importance of environment and culture in personality but emphasizes that it is the genetic, or structural, factors that first need to be identified.

The NPA model acknowledges that the genetic bases of personality are themselves complex. It assumes at least four tiers to this genetic basis:

- male or female gender
- character type based on the three NPA traits
- temperament, or the degree of activity or excitability of an individual in the Pavlovian sense
- other facets of personality, such as Raymond Cattell's 16 Personality Factors or Hans Eysenck's P-E-N model of personality.

The NPA model, thus, focuses on only the second of these four tiers, acknowledging that temperament and other facets of personality may involve a large number of genes.

Traits of sanguinity (N), perfectionism (P) and aggression (A)

Karen Horney (Fig. A2) advanced the concept that at maturity there exist at least three expansive character types, namely the "narcissistic," the "perfectionistic" and the "arrogant-vindictive" [5]. Extending these ideas, the NPA model posits that the human character rests primarily on the existence of three major traits: sanguinity (N), perfectionism (P) and aggression (A). Each of these traits is assumed to exist as the expression of a single major pleiotropic gene. Horney considered that the traits have environmental origins, being the result of an individual's desperate search for dominance in the context of a stifling upbringing [5]. The NPA model — in ascribing the traits to genetic origins — emphasizes biological attributes associated with the traits.

Aggression

The behavioral trait of aggression is proposed to be the most labile of the three [6]. The stereotypic acts associated with this trait involve body posturing, gestures, and eye contact of intimidation and deference, with individuals having this trait continually competing with each other on a scale of dominance and submission. The trait of aggression corresponds to a striving for *power* over one's environment, hence it is one main component of competitiveness in social relations, or ambition. In a pejorative connotation the trait may reveal itself in the context of sadism or sadomasochism. The facial expression is non-sanguine, i.e., tending toward sallowness or pallor in individuals of light skin color. The hallmark of the trait of aggression is a mass discharge of the sympathetic nervous system: the "flight or fight" response or the aggressive-vindictive rage. During the expression of this rage, the facial complexion of pallor is accentuated.

Fig. A2. Karen Horney (1885-1952)

Sanguinity

The trait of sanguinity (Horney's "narcissism") is proposed to be less labile than that of aggression (where individuals may be constantly altering their character states on a scale of dominance and submission) [6]. The stereotypic acts associated with the trait include self-flaunting body posturing, expansive arm gestures, bowing, instinctive self-adornment, and a natural attraction to the limelight of personal recognition. Individuals having only this trait (of the three) are competitive but non-aggressive in their strivings for recognition. The trait corresponds to a striving for *glory* in one's environment, representing the second main component of human ambition. In the absence of mediating factors, the unbridled trait of sanguinity may reveal itself in the context of conceit, exhibitionism, vanity or messianism. An associated facial expression includes the radiant gingival smile (broadly exposing the gums and teeth). The facial complexion in individuals of light skin color tends toward blood-red or ruddy. Hallmarks of the trait include blushing, flushing, and a mass discharge of the para-sympathetic nervous system: the narcissistic rage of defense and withdrawal. During expression of this rage the normally sanguine complexion becomes even more florid.

Perfectionism

The trait of perfectionism in the NPA model is not a basic drive of ambition and is not associated with a rage reaction [6]. Rather it is a mediator of the unbridled drives of aggression and/or sanguinity. The stereotypic acts associated with the trait of perfectionism are obsessiveness, compulsiveness, repetition, and the maintenance of neatness, order and symmetry. A clue to the nature of the trait lies in the compulsive, repetitive mannerisms of autistic children and some adult schizophrenic individuals. The behavioral pattern is often ritualistic and the speech characterized by echolalia. It is posited that such autistic and schizophrenic individuals are those in whom the two components of ambition, i.e., aggression and sanguinity, have been suppressed by genetic or environmental factors, either congenitally, in childhood, or after maturity, thus revealing in the individual a primitive state of perfectionism.

Character types

The notion that humans exhibit only a limited number of discrete character types can be traced back to the time of the ancient Greeks, in particular to the theory of humors (blood, black bile, yellow bile and phlegm: Fig. A3) The NPA model attempts to relate genetic NPA types to these character types of antiquity, as well as to the classic personality disorders of modern psychiatry.

Fig. A3. Character types of the ancient theory of humors: *Phlegmaticus, Cholericus, Sanguineus* and *Melancholicus*. [*J.K. Lavater, ca. 1775*]

Dominance: Dominant character types

In **Dominant types** the traits A and N, if present at all, are fully expressed [6]. The NPA model generates the following character types:

N type

The *sanguine (N) type* is found in the writings of Karen Horney [7] and others who have developed the classic psychiatric views of narcissism. In the NPA model this type is the equivalent of the sanguine character type described by the ancients. The important attributes of this type are: expansiveness but unaggressiveness, non-perfectionism, a tendency to flamboyant self-adornment, a natural attraction to the limelight, the gingival smile of recognition and the florid narcissistic rage. In extreme forms this type appears as a self-anointed visionary, a proselytizing evangelist or a messianic personality.

A type

The *aggressive (A) type* corresponds to Horney's arrogant-vindictive type and to her concept of "moving against people" [8]. In the NPA model this is the classic choleric character type of antiquity. The main attributes of this type are: unbridled arrogance, instinctual vindictiveness, non-perfectionism, no tendency to self-adornment, a wry or sardonic grin in place of a gingival smile, and the pallid-complexioned aggressive-vindictive rage. In extreme forms this type appears as a sadistic personality, as an extroverted paranoid personality, or as the so-called antisocial or sociopathic personality.

NA type

The *sanguine-aggressive (NA) type* is regarded to be a composite of the previously described sanguine and aggressive types. Horney described the essence of this character type, in the female, in an article, "The overvaluation of love: a study of a common present day type" [9]. The main attributes of this type are: a sanguine complexion, synergistic merging of unbridled narcissism and aggression, hyperactivity, non-perfectionism, a tendency toward extreme self-adornment, exhibitionism in the limelight, a "flashy" extroverted smile, a tendency toward hypersexuality, and the capacity to exhibit the narcissistic, aggressive-vindictive or combined narcissistic-aggressive rages. In extreme forms this type appears as the hypomanic, histrionic or hysterical personality.

NP type

The attributes of the *sanguine-perfectionist (NP) type* were described by Horney in her exposition of the "perfectionist type" [4]. In the NPA model this encompasses the classic phlegmatic type known to the ancients. The main qualities of this type are: a tendency toward a sanguine complexion, industriousness, orderliness, an intense sense of duty, unaggressiveness, stubbornness, negativism, a tendency to ruminate, perfectionistic rather than unbridled self-adornment, an uncommonly seen gingival smile of recognition, and the capacity to exhibit the florid narcissistic rage. In extreme forms this character appears as the obsessive-compulsive personality.

PA type

The *perfectionistic-aggressive (PA) type* is alluded to by Horney in her mention of aggressive types who function in the capacity of a "power behind the throne" [8], that is, personages who utilize intellectual qualities and planning rather than overt aggression to achieve their aims. In the NPA model this is the classic non-sanguine, austere melancholic personality of the ancients. The principal qualities of this type are: a non-sanguine complexion, passive aggressiveness, dour perfectionism, vigilance, manipulativeness, a proud bearing, haughty reservedness, a calculated vindictiveness, a lack of an innate tendency to self-adornment, a sardonic grin, and the pallid-complexioned aggressive-vindictive rage. In extreme forms this is the passive-aggressive, rebellious-distrustful, or ruminating paranoid personality.

NPA type

The *sanguine-perfectionistic-aggressive (NPA) type* was not explicitly described by Horney, although she did note that the three traits can coexist in the same individual [10]. The main attributes of this type are: a sanguine complexion, a loud voice, dynamism with a tendency to be overbearing, bombastic garrulity, intense eye contact, a strong sense of duty, a bent toward conventional values, unpretentious self-adornment, an outgoing smile of moderate intensity, and the capacity to exhibit the narcissistic, aggressive, or explosive narcissistic-aggressive rages. In the extreme cases this individual is the managerial-autocratic or explosive personality.

Submission: Passive Aggressive character types

In Passive Aggressive types the trait of aggression is not fully expressed [6]. The NPA model defines two gradations of relative submission: *non-compliance*, in which the individual is basically submissive but is easily activated to an energetic state of aggression, and *compliance*, in which the individual tends to remain in a profound state of submission.

In the model the state of repressed aggression most often has a genetic basis, the result of a congenital, inherited, incomplete expression of the gene for the trait A. However, the model also allows for environmental causes, the state of submission being induced during the juvenile period on the basis of environmental constraints to character development. That is, phenocopies (based on environmental factors) of a genetically disposed submissive state may exist. Also, like Dominant types having full expression of the trait A, Passive Aggressive types may exhibit the aggressive A rage.

Non-compliant types

The model denotes the state of non-compliance by A–, obtaining the following *non-compliant* phenotypes:

- **Aggressive (A–)**
- **Perfectionistic-aggressive (PA–)**
- **Sanguine (NA–)**
- **Sanguine-perfectionistic (NPA–)**

Compliant types

The model denotes the state of compliance by A=, obtaining the following *compliant* phenotypes:

- **Aggressive (A=)**
- **Perfectionistic-aggressive (PA=)**
- **Sanguine (NA=)**
- **Sanguine-perfectionistic (NPA=)**

The *NPA– non-compliant type* above corresponds to active, motivated, non-confrontational individuals whose baseline personality tends toward submissiveness, as described by Horney in her discussion of inverted sadistic behavior [11]. In the therapeutic setting, these individuals are found over the spectrum of the "Type A," dependent, and phobic-anxious personality. The *NA– type* is a non-perfectionistic, active individual exhibiting pronounced narcissistic behavior. In the therapeutic setting this is a cyclothymic or dependent histrionic personality.

The *compliant types NA=* and *NPA=* above correspond to more profoundly submissive individuals, having more pronounced tendencies toward masochistic behavior [12]. They correspond to Karen Horney's compliant "self-effacing" personality and to her concept of "moving toward people" [13].

Resignation: Resigned character types

In the character state of resignation the trait of aggression is stunted after maturity because of environmental constraints [6]. Unlike the Passive Aggressive types who readily involve themselves in the relative competition of dominance and submission (and sometimes sadomasochism), Resigned types remain relatively detached from such activities and only with difficulty can be stressed to a state of active aggression. However, like Passive Aggressive types, Resigned types can be induced into the aggressive A rage.

The model denotes the state of resignation by –A, obtaining the following Resigned phenotypes:

- **Aggressive (–A)**
- **Perfectionistic-aggressive (P–A)**
- **Sanguine (N–A)**
- **Sanguine-perfectionistic (NP–A)**

The sanguine Resigned types, having the N trait, correspond to detached individuals, as described by Horney. She considered that "moving away from people" was a maladaptive response that could develop as a growing individual struggled toward maturity [14]. The *NP–A type* would tend to have strong perfectionistic tendencies, while the *N–A type* would be more labile.

Borderline types and mental illness

In the NPA model, Borderline types possess only one of the traits of ambition (N or A) and it is only partially expressed. Types in which both traits (N and A) are either absent or profoundly suppressed fall into categories of mental illness, in particular schizophrenia [6]. Thus, NPA theory predicts that the categories of borderline personality and schizophrenia are heterogeneous, depending on the underlying NPA character structure. Examples of Borderline types would be the A– or PA– types above. Types falling into the categories of mental illness would be the compliant Passive Aggressive types, A= or PA=.

One aspect of the model focuses on the Dominant types N and NP, which lack the trait A [6]. In analogy with partial expression of the trait A, the theory identifies states of incomplete expression of the trait N, denoted as N–, N= and –N. Examples of Borderline types would be N– or N–P types. Types falling into the categories of mental illness would be N= or N=P, the latter being a perfectionistic, autistic individual.

Dominance and submission

In the NPA model, Dominant character types having the trait A have the potential of being reduced to a subdued state acutely or to a subjugated state chronically (see the figure on p. 45). Similarly, non-compliant Passive Aggressive types have the potential of being activated to an energetic A+ state resembling dominance, usually for short periods of time. Thus, the model emphasizes the potential lability of trait A in social relations, with Dominant and Passive Aggressive types continually altering their behavior in competitive interactions with other individuals and in the context of mating. In the extreme, some of these relationships fall into the category of sadomasochism [15]. Resigned types, in their detachment from social interactions, usually avoid dominance-submission relationships and, in particular, hierarchal structures where "pecking orders" predominate.

Mendelian transmission of NPA traits

On the basis of archetypal examples, the model assumes that in their full expression the NPA traits are transmitted by autosomal genes, with traits A and N being recessive and trait P being transmitted in the dominant mode [6]. The alleles corresponding to full expression and total suppression of the trait A are denoted by **a** and A_0, respectively, and the corresponding alleles for the trait N are denoted by **n** and N_0. For the trait P two alleles **P** and p_0 are posited, corresponding to full expression or total absence of the trait P, on the assumption that the trait is transmitted with complete penetrance. This scheme of inheritance is consistent with the notion that the alleles A_0 and N_0 control the production of inhibitors of the traits A and N at the level of the central nervous system, with alleles A_0 and N_0 being dominant with respect to **a** and **n**.

The scheme leads directly to Table A1 below, showing the possible phenotypes of progeny according to the phenotypes of the parents:

	N	A	NP	NA	PA	NPA
N	N -- -- NA -- -- -- -- --	"	"	"	"	"
A	N -- -- NA -- -- 0 A	-- -- -- NA -- -- -- A	"	"	"	"
NP	N NP -- NA NPA -- -- --	N NP P NA NPA PA 0 A	N NP -- NA NPA -- -- --	"	"	"
NA	N -- -- NA -- -- -- --	-- -- -- NA -- -- -- A	N NP -- NA NPA -- -- --	NA -- -- -- --	"	"
PA	N NP P NA NPA PA 0 A	-- -- -- NA NPA PA -- A	N NP P NA NPA PA 0 A	-- -- -- NA NPA PA -- A	-- -- -- NA NPA PA -- A	"
NPA	N NP -- NA NPA -- -- --	-- -- -- NA NPA PA -- A	N NP -- NA NPA -- -- --	-- -- -- NA NPA -- -- --	-- -- -- NA NPA PA -- A	-- -- -- NA NPA -- -- --
FATHER OR MOTHER	N	A	NP	NA	PA	NPA

Table A1. Possible phenotypes of children according to the phenotypes of the parents. The phenotypes of the father and mother are shown along the axes of the table. The P and null (0) phenotypes by the model are non-viable and would result in miscarriage, stillbirth or an infant who fails to thrive.

The table shows:
- N and A individuals need not have N or A parents. Such individuals can arise *de novo* so long as at least one of the parents is an NP and PA individual, respectively.
- PA individuals must have at least one parent who is of either the PA or A type.
- NP individuals must have at least one parent who is of either the NP or N type.
- NA individuals can arise *de novo* from any combination of phenotypes.
- The mating of two NA types can yield progeny of only NA types.

- The mating of an NPA type with an NA type can yield progeny of only NPA or NA types.

- Certain combinations of parental genotypes may lead to zygotes having only the P trait (P phenotype) or lacking all three traits (null phenotype, denoted by 0). According to NPA theory, zygotes of P or null phenotype would be non-viable. Thus, the model predicts partial or complete infertility in some combinations of parental phenotypes, these being N×A, N×PA, NP×A and NP×PA.

Implications of a trait theory based on genetics
Population Genetics

A trait theory based on genetics would imply that the personality structure of a population could be expressed in definitive mathematical terms. The NPA model is amenable to the Hardy-Weinberg approach to quantify the distribution of NPA character types in a given subpopulation [16]. With the usual assumptions of gene frequencies n, p and a and random mating, incidences of Dominant character types are given in Table A2, below. Because of the occurrence of non-viable P and null (0) phenotypes, the assumptions of Hardy-Weinberg equilibrium would not be strictly valid: the incidences generated by the expressions in Table A2 below represent the phenotypes of the first generation only.

The assumption of numerical values for the three gene frequencies n, p and a generates a hypothetical subpopulation, or habitancy [16]. In Table A3 below six habitancies are given with descriptive labels: *Polymorphic,* (or "Balanced"), *Punctilious, Sublime, Demonstrative, Authoritarian* and *Militant*. The intent of the labels is to emphasize the very different tenors of each of the distributions of character types.

Table A3 demonstrates that:

- Relatively small changes in gene frequencies could cause large changes in the phenotype frequencies.

- The frequencies of non-viable P and null types are significant for some of the habitancies, on the order of 1 to 9 percent.

→

APPENDIX

Relative incidence of phenotypes on basis of gene frequencies n, p and a

Phenotype	Relative incidence
N	$n^2 \times (1-p)^2 \times (1-a^2)$
A	$(1-n^2) \times (1-p)^2 \times a^2$
NP	$n^2 \times p(2-p) \times (1-a^2)$
NA	$n^2 \times (1-p)^2 \times a^2$
PA	$(1-n^2) \times p(2-p) \times a^2$
NPA	$n^2 \times p(2-p) \times a^2$
P	$2n(1-n) \times p(2-p) \times 2a(1-a)$
null (0)	$2n(1-n) \times (1-p)^2 \times 2a(1-a)$

Table A2. Relative incidences of phenotypes for the first generation. The incidence for each phenotype is the product of three probabilities, corresponding to the presence or absence of the three traits N, P and A. The P and null types are non-viable and contribute neither to parentage nor issue.

	HABITANCY					
Phenotype	Balanced	Punctilious	Sublime	Demonstrative	Authoritarian	Militant
N	7	3	77	2	1	1
A	3	<1	<1	2	17	34
NP	22	78	18	7	2	1
NA	13	<1	3	20	6	11
PA	9	2	<1	7	52	35
NPA	39	8	1	61	17	12
P	4	8	<1	1	4	2
null (0)	1	<1	1	<1	1	2
Gene frequencies	n = 0.90 p = 0.50 a = 0.80	n = 0.90 p = 0.80 a = 0.30	n = 0.99 p = 0.10 a = 0.20	n = 0.95 p = 0.50 a = 0.95	n = 0.50 p = 0.50 a = 0.95	n = 0.50 p = 0.30 a = 0.95

Table A3. Frequencies of phenotypes in six habitancies (per 100 zygotes, or pregnancies). The P and null (0) phenotypes are non-viable. Non-viable types arise when the zygote has neither trait N nor A. The above analysis is confined to Dominant character types on the assumption of two alleles for each NPA gene.

Evolutionary origins of NPA traits

The assumption of a genetic basis for the traits N, P and A implies that their origins reside in the evolution of humans from precursor species, and in particular, that the traits are likely to be found in primates other than *Homo sapiens*. As examples, the model leads to proposed character types as follows:

- The omnivorous, hierarchal, unsmiling olive baboon, known for its lengthy grooming rituals, would be a likely perfectionist-aggressive PA type.

- The herbivorous, aloof, phlegmatic orangutan and gorilla, capable of gingival smiles, would be likely NP types.

- Akin to humans, the omnivorous, promiscuous chimpanzee, also capable of the gingival smile, would likely have a heterogeneous distribution of types, with NA and NPA types predominating.

Fig. A4. NPA theory proposes that the olive baboon is a likely perfectionist-aggressive PA type.

Predictive aspects of NPA model

The model would have the potential to be predictive in the following categories:

- The possible genetic character types of children could be deduced from the character types of parents.
- Relations could be defined between genetic character type and susceptibility to certain physical and mental diseases.
- Combinations of parental character types prone to infertility problems (miscarriage and stillbirth) could be identified, these combinations being ones which permit the occurrence of a fetus having neither trait N nor A.
- Allele frequencies for the NPA traits, as well as the resultant distributions of NPA character types, in various societies could be analyzed on the basis of well-known principles of population genetics.
- Studies with primates could confirm a biological basis for behavior in the areas of sociobiology and evolutionary psychology.

Criticism and controversy

Controversy has always followed past positions taken by the scientific community relating human behavior to inheritance, as in Arthur Jensen's theories of intelligence, Herrnstein and Murray's "The Bell Curve," or Lewontin and colleagues' "Not in Our Genes." The NPA personality theory is not exempt. The result of the "nature versus nurture" debate has been that a gauntlet had been thrown to those who espouse genetic underpinnings to behavior: "show us the relevant genes."

The slow progress of unraveling of the genetic basis of personality is the subject of a recent review article by Jang and colleagues [2]. They point out the lack of any genetic framework in the classification of the Diagnostics and Statistical Manual of American psychiatry (DSM-IV), and the pressing need to identify "genetically crisp" characteristics — or genetic traits of behavior that are independent of competing genetic and environmental influences.

The NPA model posits sanguinity to be a genetic trait, being related to the parasympathetic branch of the autonomic nervous system, just as aggression is classically related to the sympathetic branch. This concept of sanguinity, and the associated narcissistic rage, is not found in any branch of classical medicine or psychiatry and remains a key point requiring validation. Of note is the recent study by Livesley and colleagues [3] with identical and fraternal twins. They found that of a total of eighteen dimensions of personality it was narcissism that had the highest heritability.

The manuscript of the NPA model was copyrighted with the Library of Congress in 1982, being published in book form in 1985 [17] and in a peer-reviewed journal in 1990 [6]. A revised electronic edition in PDF format was released in 2004 and the online NPA personality test in 2005. Studies are in progress utilizing the NPA personality test in obstetric and gynecological
patients [18].

Although the NPA model is several decades old, it has not been validated in the sense of withstanding scrutiny by the scientific method — — as is true of all other theories of personality as well. Given the recent advances in deciphering the human genome, such scrutiny may soon be possible. The ideas of Karen Horney have been resilient over time, and the validity of her observations that form the basis of the NPA model awaits the relevant studies in the realm of behavioral genetics.

References

Benis, A.M. *Toward Self and Sanity: On the genetic origins of the human character*, Psychological Dimensions, New York, 1985. ISBN 0884370747 [2nd edition, 2008. ISBN 9780615262147]

Benis, A.M. and J.H. Rand (1986). A model of human personality based on Mendelian genetics (abstract). *Proceedings of the American Association for the Advancement of Science,* Publication 86-5, 124.

Benis, A.M. (1990). A theory of personality traits leads to a genetic model for borderline types and schizophrenia. *Speculations in Science and Technology 13* (3), 167-175.

Freud, Sigmund. "Heredity and the aetiology of the neuroses," in *Early Psycho-analytic Publications,* Hogarth, London, [1896] 1962.

Horney, Karen. *Neurosis and Human Growth*, Norton, 1950.

Horney, Karen. *Our Inner Conflicts*, Norton, 1945.

Horney, Karen. *New Ways in Psychoanalysis*, Norton, 1939.

Horney, Karen. *Feminine Psychology*, Norton, [1922 to 1937] 1967.

Jang, K.L., Vernon, P.A. and W.J. Livesley (2001). Behavioural-genetic perspectives on personality function. *Canadian Journal of Psychiatry 46*, 234-244.

Livesley, W.J., Jang, K.L., Jackson, D.N. and P.A. Vernon (1993). Genetic and environmental contributions to dimensions of personality disorder. *American Journal of Psychiatry 150*, 1826-1831.

Stone, Michael H. *The Borderline Syndromes*, McGraw-Hill, 1980.

Citations

1. *Personality*, in Wikipedia.
2. Jang *et al.* (2001). Behavioural-genetic perspectives.
3. Livesley *et al.* (1993). Genetic and environmental contributions.
4. Horney, *Neurosis and Human Growth*, Chapter 8: The expansive solutions: the appeal of mastery.
5. Horney, *Neurosis and Human Growth*, Chapter 4: Neurotic pride.
6. Benis (1990). Theory of personality traits leads to genetic model.
7. Horney, *New Ways in Psychoanalysis*, Chapter 5: The concept of narcissism.
8. Horney, *Our Inner Conflicts*, Chapter 4: Moving against people.
9. Horney, *Feminine Psychology*, pp. 182-213.
10. Horney, *New Ways in Psychoanalysis*, p. 97.
11. Horney, *Our Inner Conflicts*, Chapter 12: Sadistic trends.
12. Horney, *New Ways in Psychoanalysis*, Chapter 15: Masochistic phenomena.
13. Horney, *Our Inner Conflicts*, Chapter 3: Moving toward people.
14. Horney, *Our Inner Conflicts*, Chapter 5: Moving away from people.
15. Horney, *Neurosis and Human Growth*, Chapter 10: Morbid dependency.
16. Benis, *Toward Self and Sanity*, Chapter 10: Genetics.
17. Benis, *Toward Self and Sanity*.
18. by Donna K. Hobgood, M.D., Clinical Attending Physician, University of Tennessee College of Medicine, Chattanooga.

Illustrations

Karen Horney: "Studio photo" courtesy of Karen Horney Papers, Manuscripts and Archives, Yale University Library, New Haven. Copyright unknown.

Character types according to theory of humors: From Johann Kaspar Lavater, *Physiognomics*, ca. 1775.

Olive baboon: U.S. Fish and Wildlife Service.

Source

This article originally appeared in *Wikipedia*, the online encyclopedia in May 2006. It was later deleted for reasons of non-notability. The reference was: "NPA personality theory," *Wikipedia, The Free Encyclopedia,* 2 July 2006, Wikimedia Foundation:

http://en.wikipedia.org/wiki/NPA_personality_theory.

GLOSSARY

aggression, trait of The genetically determined set of behavioral attributes that form the basis of man's desire to survive by maintaining a position of power over his environment. Trait A of the model.

aggressive rage (A rage) Mass discharge of the sympathetic nervous system related to the A trait of aggression. See *narcissistic rage*.

allele An alternative form of a gene at a given locus.

antisocial personality disorder A personality disorder characterized by aggression and a disregard for other people.

association The occurrence together, in a family or population, of two characteristics in a frequency greater than that predicted by chance. The association may or may not reside in the expression of linked genes.

assortative mating Non-random or preferential mating among individuals of a population.

autosomal Pertaining to a non-sex chromosome.

autistic spectrum Developmental disorders characterized by restricted and repetitive behavior that impair social interaction and communication.

blushing Vasodilation in an emotional context, in the skin of the face, neck and upper chest. According to the model, individuals having the N trait have an increased predisposition to blushing and flushing.

Borderline type An NPA type in which neither trait N nor A is fully expressed.

breed true A trait is said to breed true if two parents of the same phenotype always produce offspring of that same phenotype exclusively. The NA type is the only type of the model that always breeds true: any two NA types can have only NA offspring.

carrier An individual who carries a gene that is not expressed.

chromosomes The cell structures containing the genetic material DNA. The human genome is composed of 46 chromosomes: 22 pairs of autosomes and 2 sex chromosomes.

cognition The acts of thinking, feeling, knowing, reasoning and learning, including both awareness and judgment.

complementary genes Genes that produce different phenotypic effects depending on whether they are present separately or together. In the model, the complementary genes A_0 and N_0 — corresponding to the inhibition of traits A and N — produce a lethal effect in a zygote.

Dobzhansky-Muller Refers to the Bateson-Dobzhansky-Muller model of the genetic incompatibility of hybrid types, as it applies to speciation in evolutionary theory.

dominant trait Refers to Mendelian dominance. Not to be confused with *Dominant type*.

Dominant type An NPA type in which the traits N and/or A are fully expressed. The six types are: N, A, NA, NP, PA and NPA.

epistasis The condition in which a gene at one locus suppresses the expression of a gene at another locus.

exhibitionism Tendency toward display or extravagant behavior. Exhibitionism is most often a manifestation of the unbridled N trait.

expressivity The degree to which a genetic trait is observed in the phenotype. Variable expressivity may be caused by modifier genes or by environmental effects.

extrovert An individual whose attention and interests are directed primarily toward others.

failure to thrive Referring to an infant who does not develop normally and eventually succumbs. According to the model, certain parental combinations of NPA types will be prone to miscarriages, stillbirths, and infants who fail to thrive.

"fight-or-fight" reaction Behavioral response associated with mass discharge of the *sympathetic nervous system*. See also: *aggressive rage*.

gene A fundamental unit of heredity, composed mainly of DNA. Genes are arranged in linear order on the chromosomes and determine the *genotype* of the individual.

gene frequency The probability of an allele's existing at a given chromosomal locus in an individual of a given population.

genotype The genetic constitution of alleles in an individual with respect to a gene locus or loci. See *phenotype*.

gingival smile A broad smile, revealing the gums of the upper teeth, related to the N trait.

habitancy   In NPA population genetics, the inhabitants of a region, taken collectively, or a subpopulation. For ease of communication we define the following habitancies:

Polymorphic — a mixture of NPA types
Punctilious — mainly NP types
Sublime — mainly N types
Demonstrative — mainly NPA types
Corybantic — mainly NA types
Authoritarian — mainly PA types
Militant — mainly A types

The *Authoritarian* and *Militant* habitancies have mainly non-sanguine types, the others being primarily sanguine. Some examples are: *Polymorphic*: USA, UK, South Africa, Australia; *Punctilious:* Northern Italy, Switzerland, Taiwan; *Sublime:* East Africa, Polynesian Islands; *Demonstrative:* Southern Italy and France, Northern Iran; *Corybantic:* Brazil, Senegal, Aboriginal central Australia; *Authoritarian:* Western Russia, Eastern Europe; *Militant:* Yemen, Arab region of Iraq.

Hardy-Weinberg principle   The principle states that allele and genotype frequencies in a population will remain constant from generation to generation in the absence of other evolutionary influences. In the present model, infertile parental pairs would mate, producing non-viable P and null types. Hence, in order for the proportions of the genotypes of the population to remain constant, the occurrence of assortative mating or reproductive advantages in certain NPA types would have to be invoked.

heterozygous   Having dissimilar alleles at a locus of a homologous pair of chromosomes.

homozygous   Having similarly functioning alleles at a locus of a homologous pair of chromosomes.

Horney, Karen (1885–1952)   German-American psychiatrist of Dutch and Norwegian heritage.

hybrid   An individual whose parents belong to two different varieties of a species, or to two different species.

introvert   An individual whose interests are predominantly concerned with his own mental life.

lethal gene   A gene that renders non-viable an organism or cell possessing it. According to the present model, the genes corresponding to the absence of traits N and A, when present together, act as *complementary genes* to produce a lethal effect (a non-viable zygote).

modifier genes   Genes that modify an observed physical or behavioral trait.

mutation   A change in the DNA structure of a gene. If the change occurs in a gamete (reproductive cell), then the alteration may be perpetuated in subsequent generations.

narcissism   From Narcissus, the figure in Greek mythology who fell in love with his own reflected image. In the present model, narcissism is related to the *unbridled* N trait of sanguinity.

narcissistic personality disorder (NPD)   Subjects diagnosed with NPD will likely be individuals having the *unbridled* N trait.

narcissistic rage (N rage)   Mass discharge of the autonomic nervous system related to the N trait of sanguinity.

non-sanguine   Refers to NPA types that lack the trait N.

NPA+ type   For clarity, this designation refers specifically to the NPA Dominant type of the model, as distinguished from the generic NPA type.

null zygote   According to the model, a *zygote* in which none of the NPA traits are expressed. Null zygotes do not develop into viable individuals.

Passive Aggressive type   An NPA type in which trait A is genetically partially inhibited.

penetrance   The expression of a trait when the genotype is present. Thus, in "incomplete penetrance" a certain proportion of individuals will not exhibit the trait although the appropriate genotype is present.

perfectionism   The P trait of the model, appearing in behavior as 1) the achievement of order by persistence and repetition, and 2) as a trait that modulates the expression of the unbridled N and A traits.

personality   A collection of behavioral patterns unique to an individual that is consistent over time.

phenotype   The observable traits in an individual. The NPA character types (N, NP, PA, etc.) are phenotypes.

pleiotropism   The determination of multiple somatic and/or behavioral characteristics by a single gene.

polygenic   Referring to the influence of several genes determining the expression of a trait.

recessive trait   Refers to a trait that is expressed only when the causative gene is present in the *homozygous* state, i.e., similarly functioning alleles on both chromosomes of an autosomal pair. See *dominant trait*.

Resigned type   An NPA type in which trait A is partially inhibited due to environmental stress after maturity.

sanguine, sanguinity   According to ancient physiology, belonging to one of the four "temperaments" in which blood predominates over the other three humors, leading to a ruddy countenance and exuberant behavior. In the NPA model, a sanguine character type is any type having the N trait. See *non-sanguine*.

schizophrenia   A group of psychotic disorders characterized by disturbances in thought, mood and behavior. In the present model, a kind of schizophrenia may occur in an individual if there is a lack of normal expression of both the N and A traits.

temperament   The general level of activity, reactivity or excitability of an individual in the Pavlovian sense.

unbridled trait   The presence of fully expressed trait N or A without modulation by the P trait.

zygote   A fertilized egg that may develop into a fetus.

REFERENCES & NOTES

Chapter 1: Introduction: Sanguinity and Aggression
[1] Wahlsten (2012).

[2] Benis A.M. (2017a).

Chapter 2: The NPA Model
[3] See Appendix: *Synopsis of NPA Personality Theory*. This article, originally published in Wikipedia, is a concise summary of the NPA model. In Horney's last book, *Neurosis and Human Growth* (1950), she presents her concept of three "expansive types": "the narcissistic, the perfectionistic and the arrogant-vindictive".

[4] Benis (1985/2017). The gingival smile is seen also in other Primates, including the great apes and some Old World monkeys, implying that it has deep evolutionary roots.

[5] The caricatures were originally published in Benis (1985/2017). They have been updated in a monograph: Benis (2017b): *Caricatures of the NPA Personality Types*.

[6] Benis A.M. (2017c). Chap. 7: Typing people.

Chapter 3: Genotypes and infertility
[7] Benis (1985/2017). Chap. 10: Genetics.

Chapter 4: Population genetics
[8] Benis A.M. (2017a).

[9] Orr (1995, 1995a).

[10] Dobzhansky (1970).

[11] Benis (1985/2017). For an updated monograph, see Benis (2017a).

[12] Alvarez (2005).

[13] Benis (1985/2017). For an updated monograph, see Benis (2017a).

[14] There are areas in the world where a Punctilious habitancy (non-aggressive types) is adjacent to an Authoritarian habitancy (non-sanguine types). See Benis (2017a).

[Chapter 5: Case-control studies
[15] The NPA genes are posited to be major genes that determine the basic personality structure. If this is true, then the NPA traits may be related to a variety of conditions, syndromes and diseases. See Benis (1985/2017).

[16] Benis A.M. (2018): *The Enigma of Short Parents Who Have Tall Children*.

[17] Benis A.M. (1990); Benis A.M. and D. K. Hobgood (2011). See also the *Appendix:* Borderline types and Mental illness.

BIBLIOGRAPHY

Alvarez, L. (2005). Narcissism guides mate selection: Humans mate assortatively, as revealed by facial resemblance, following an algorithm of "self seeking like." *Evolutionary Psychology 2*, 177-94.

Benis A.M. (1985). *Toward Self & Sanity: On the Genetic Origins of the Human Character,* Psychological Dimensions, New York. Revised edition 2017, as *NPA Theory of Personality*, New York, KDP/Amazon, ISBN 978-1521283295.

Benis, A.M. (1990). A theory of personality traits leads to a genetic model for borderline types and schizophrenia. *Speculations in Science and Technology 13* (3), 167-175.

Benis A.M. and D. K. Hobgood (2011): Dopamine receptor DRD3 codes for trait aggression as Mendelian recessive, *Medical Hypotheses* 77(6):1108-10.

Benis A.M. (2017a). *Geographic Distribution of Genetic Character Traits Based on the NPA Theory of Personality*, KDP/Amazon, ISBN 978-1520430317.

Benis A.M. (2017b). *Caricatures of the NPA Personality Types*, KDP/Amazon, ISBN 978-1520966977.

Benis A.M. (2017c). *How Your Personality Type Is Inherited: The NPA Model of Genetic Traits*, KDP/Amazon. ISBN 978-1387232543.

Benis A.M. (2017). *NPA Personality Theory: The Essentials,* KDP/Amazon, ISBN 978-1521399910.

Benis A.M. (2018). *The Enigma of Short Parents Who Have Tall Children.* KDP/Amazon, ISBN 978-1983050138.

Dobzhansky T. (1970). *Genetics of the Evolutionary Process*. Columbia University Press, New York.

Horney K. (1950). *Neurosis and Human Growth*, Norton, New York.

Orr H.A. (1995). The population genetics of speciation: The evolution of hybrid incompatibilities, *Genetics* 139, 1805-13.

Orr H.A. (1995a). Dobzhansky, Bateson and the genetics of speciation, *Genetics* 144, 1331-35.

Wahlsten D. (2012). The hunt for gene effects pertinent to behavioral traits and psychiatric disorders: From mouse to human, *Dev Psychobiol* 54, 475-92.

SOURCES OF ILLUSTRATIONS

Figure 1, p. x:

The four characters of man, from Johann Kaspar Lavater (*ca.* 1775): *Physiognomics*.

Figure 2, p. 6:

Gingival smile, by max_thinks_sees. Creative Commons license via: flickr.com/photos/hundreds/2830576097.

Figure 3, p. 8:

By Anthony Moore, from Desmond Morris (1977): *Manwatching*. Courtesy of Elsevier/Equinox Publishing Projects, Oxford.

Cover

Diagram of the Mind of the Microcosm, from Robert Fludd (*ca.* 1617): *Utriusque cosmi maioris scilicet et minoris metaphysica…*

ACKNOWLEDGEMENTS

With thanks to my colleagues, D.K. Hobgood, M.D. for her continuing interest and helpful comments, and to J.H. Rand, M.D., who provided invaluable assistance with the original version of the NPA model and helped to guide the manuscript to the publisher.

ABOUT THE AUTHOR

The author received the degree of Doctor of Science from MIT. His medical training was at the Mount Sinai Medical Center in New York, where he served afterward for many years as Research Associate Professor and Director of Cardiothoracic Intensive Care. He is the author of a number of research papers and review articles.